COMPARING
ENVIRONMENTAL RISKS

COMPARING
ENVIRONMENTAL RISKS

TOOLS FOR SETTING GOVERNMENT PRIORITIES

Edited by J. Clarence Davies

RESOURCES FOR THE FUTURE
Washington, DC

Printed in the United States of America

Published by Resources for the Future
1616 P Street, NW, Washington, DC 20036–1400

Library of Congress Cataloging-in-Publication Data

Comparing environmental risks : tools for setting government
priorities / edited by J. Clarence Davies.
    p.   cm.
    Includes bibliographical references and index.
    ISBN 0-915707-79-9
    1. Environmental risk assessment.   I. Davies, J. Clarence.
GE145.C65   1995
363.1'056—dc20   95–44579
                                                 CIP

∞ The paper in this book meets the guidelines for permanence and durability of the Committee on Production Guidelines for Book Longevity of the Council on Library Resources.

This book is the product of the Center for Risk Management at Resources for the Future, J. Clarence Davies, director. It was copyedited by Eric Wurzbacher. The book was designed by Diane Kelly, Kelly Design, and paginated by Amie Jackowski. The cover was designed by Ingrid Gehle, Gehle Design Associates, Inc.

RESOURCES FOR THE FUTURE (RFF) is an independent nonprofit organization engaged in research and public education on natural resources and environmental issues. Its mission is to create and disseminate knowledge that helps people make better decisions about the conservation and use of their natural resources and the environment. RFF takes responsibility for the selection of subjects for study and for the appointment of fellows, as well as for their freedom of inquiry. RFF neither lobbies nor takes positions on current policy issues.

Because the work of RFF focuses on how people make use of scarce resources, its primary research discipline is economics. Supplementary research disciplines include ecology, engineering, operations research, and geography, as well as many other social sciences. Staff members pursue a wide variety of interests, including the environmental effects of transportation, environmental protection and economic development, Superfund, forest economics, recycling, environmental equity, the costs and benefits of pollution control, energy, law and economics, and quantitative risk assessment.

Acting on the conviction that good research and policy analysis must be put into service to be truly useful, RFF communicates its findings to government and industry officials, public interest advocacy groups, nonprofit organizations, academic researchers, and the press. It produces a range of publications and sponsors conferences, seminars, workshops, and briefings. Staff members write articles for journals, magazines, and newspapers, provide expert testimony, and serve on public and private advisory committees. The views they express are in all cases their own and do not represent positions held by RFF, its officers, or trustees.

Established in 1952, RFF derives its operating budget in approximately equal amounts from three sources: investment income from a reserve fund; government grants; and contributions from corporations, foundations, and individuals. (Corporate support cannot be earmarked for specific research projects.) Some 45 percent of RFF's total funding is unrestricted, which provides crucial support for its foundational research and outreach and educational operations. RFF is a publicly funded organization under Section 501(c)(3) of the Internal Revenue Code, and all contributions to its work are tax deductible.

Contents

Foreword

Mark Twain complained that while everybody talked about the weather, nobody did anything about it. Nearly the same could be said about comparative risk analysis. For the last nine months, most of the talking has been done by members of the United States Senate and the House of Representatives. Both houses of Congress have been debating changes in the way federal agencies set priorities and issue new regulations, including requirements that these agencies compare the risks they are proposing to reduce with other risks. (For example, is stratospheric ozone depletion a more serious problem than indoor air pollution?) Unfortunately, there is very little guidance about the value of such comparisons, how previous efforts have fared, and how future efforts should be structured.

The papers in this volume start to do something about that. With their origins in a workshop organized by RFF's Center for Risk Management, at the behest of the President's Office of Science and Technology Policy, the papers explain the origins of comparative risk analysis, the political context in which it is being put forward, its use at the state level in the United States, the limitations that might reasonably be imposed upon it, and the way in which both "experts" and the lay public might participate in making risk comparisons.

I would be surprised—though quite pleased—if these papers anticipated and answered all the right questions. Rather, their main value should be in sparking further debate about the utility of comparative risk analysis, as well as its shortcomings. If J. Clarence Davies and the contributors to his edited volume do no more than that, Mark Twain—and all the rest of us, as well—ought to be thankful.

Paul R. Portney
President
Resources for the Future

Preface

In 1993, the White House Office of Science and Technology Policy (OSTP) and the Office of Management and Budget (OMB) were considering how federal environmental, health, and safety agencies could more systematically and on a large scale establish priorities for their programs. The obvious tool for such priority setting was considered to be comparative risk analysis (CRA).

Under the leadership of Mark Schaefer, OSTP turned to the Center for Risk Mangement at Resources for the Future (RFF) for assistance in formulating methods for broader use of CRA. In August 1993, RFF hosted a small meeting involving OSTP, OMB, and a half-dozen CRA experts. The meeting agreed that CRA should be used more widely within the government and that it was both possible and useful to explore methods for using CRA.

RFF, with the active encouragement of OSTP, then applied to the Carnegie Corporation of New York, a private foundation, for funding to assist in the CRA effort. In January 1994, Carnegie awarded RFF $25,000 for this effort.

The Carnegie money was used for two purposes. First, an all-day meeting of thirty-six people, divided rather evenly between government agency representatives and nongovernment experts on risk analysis, was convened in February 1994. The meeting focused on methods by which federal agencies could use CRA to set program priorities. I chaired the meeting, and the participants included, in addition to the government officials and the academic risk experts, two former administrators of the U.S. Environmental Protection Agency (EPA), Douglas Costle and Lee M. Thomas.

The other use of the Carnegie money was to commission background papers for the meeting. RFF solicited papers from the leading scholars in the field, and everyone who was asked to contribute agreed to write a paper. The papers, particularly the one from the Carnegie Mellon group (see Chapter 6), formed the agenda for the meeting, and the meeting participants commented extensively on each of the papers. The papers were revised to include input from the workshop, thereby forming the basis for the chapters of this volume.

I am grateful for the financial support provided by the Carnegie Corporation of New York. General funding from EPA also enabled us to do the essential background and followup work that made this book possible. Jim Cole has been a wonderfully supportive EPA project officer, and I am indebted to him, as well as to Tom Kelly, David Gardiner, and many others in the EPA policy office.

Within Resources for the Future, Eric Wurzbacher skillfully edited the manuscript, and many useful suggestions came from Paul Portney, Richard Getrich, and Betsy Kulamer. Amie Jackowski typeset the book and provided assistance and computer support in its production. John Mankin provided able secretarial support.

J. Clarence (Terry) Davies
Director, Center for Risk Management
Resources for the Future

COMPARING
ENVIRONMENTAL RISKS

1

Comparative Risk Analysis in the 1990s: The State of the Art

J. Clarence Davies

Early in 1987 a small group of people met in the office of the administrator of the U.S. Environmental Protection Agency (EPA). The meeting had been convened to decide how to distribute a report that the agency had just completed. The discussion that day ranged broadly, including a suggestion that the report be given to Congress in a plain brown wrapper with no indication of authorship, because no report of this kind had ever been done by either EPA or any other government agency. The report, entitled *Unfinished Business*, was the first study to compare the risks addressed by the various EPA programs. Neither Administrator Lee M. Thomas nor anyone else in the room had a clear idea what the reaction to the report would be, but they chose to distribute the report because its findings were potentially important to the future of EPA and to the future course of environmental policy.

The reactions in the press and Congress to *Unfinished Business* were generally favorable, and various individuals concerned with environmental policy began thinking about the potential power of comparative risk analysis as a tool for sorting out and setting environmental priorities. Two years after the release of *Unfinished Business*, William K. Reilly, who had just been appointed EPA Administrator by President George Bush, decided to make risk-based priority setting a theme of his administration. That theme has been widely and heatedly debated since then, and it remains a central issue to much of federal environmental policy. As far as the contributors to this book are concerned, it is more a question of

J. Clarence (Terry) Davies is Director of the Center for Risk Management at Resources for the Future.

how best, not whether, to apply certain risk-based techniques to realize policy goals. Before discussing the perspectives of these writers, though, I would like to comment on how the risk paradigm in general, and comparative risk analysis in particular, evolved as a political issue in the mid-1990s, and how this book in its turn emerged as part of that development.

COMPARATIVE RISK ANALYSIS BECOMES A POLITICAL ISSUE

In the early 1990s several forces combined to put comparative risk analysis (CRA)—the relative ranking of risks—at the top of the environmental policy agenda. The budgetary squeeze at all levels of government made it more obvious than ever that not every environmental problem could be addressed—somehow priorities had to be set. Criticism of environmental programs in general and of EPA in particular focused in part on the low levels of risk posed by the problems being regulated. State and local governments chafed under federal requirements to spend significant amounts of money on problems that the states and localities considered both low risk and low priority. A number of local officials focused on CRA as a tool by which they could persuade Washington to pay attention to local priorities.

The state and local interest in CRA soon resonated in the U.S. Congress, which turned its collective attention for the first time to the complex and obscure topic of risk analysis. True, a few lonely congressional voices had previously focused on risk analysis. Congressman Don Ritter had introduced risk-based legislation as early as 1979: Senator Daniel Patrick Moynihan, since 1991, had been proposing legislation requiring EPA to periodically conduct a CRA of its programs. However, the 103rd Congress was the first to put risk analysis high on the agenda.

Because congressional understanding of risk analysis in general, and CRA in particular, was not very deep, it generally failed to distinguish either between individual risk assessments and comparative risk analyses, or between the two types of CRA: a large-scale comparison of programs (or problems) on the one hand and comparison of specific risks on the other. Thus the debates about risk incorporated business concerns about EPA's methodology for performing individual risk assessments and a general concern about government overregulation with the interest in using risk analysis as a basis for setting program priorities.

For all its confusion of risk categories, the 103rd Congress did indeed focus legislatively on the topic with the Johnston Amendment, first proposed by Senator Bennett Johnston of Louisiana as an amendment to a bill elevating EPA to cabinet status. The Johnston Amendment would have required EPA to do a benefit-cost analysis of major regulations and to compare the risk being regulated to other risks. Controversy over the amendment was sufficient to prevent passage of any major environmental legislation. The Johnston language was added to a Department of Agriculture reorganization bill that was enacted, but the enacted language applied only to Agriculture regulations, and no other legislative action on risk was enacted.

After the 1994 elections, the newly dominant Republicans in the 104th Congress made risk-oriented legislation part of the "Contract with America" that they had pledged to uphold during the campaign. Within the first three months of the legislative session the House had passed a detailed regulatory reform bill with risk analysis as a major feature. (As of mid-October 1995, the Senate had not passed comparable legislation. Senator Robert Dole, the majority leader, had made three attempts to close off debate on the regulatory reform measure but had failed each time.) Both bills contained detailed requirements for risk and cost-benefit analyses to be conducted on all major regulations, and both expanded the scope of judicial review of the analyses.

The congressional interest in risk analysis had its executive branch analog. Risk assessments had been used routinely as a decisionmaking tool in EPA. However, programmatic CRA, which was pioneered by EPA, had not been done by any other agency and was not a routine procedure at all within EPA. The President's Office of Science and Technology Policy (OSTP) had a strong interest in seeing CRA used more widely to set priorities.

Under the leadership of Mark Schaefer, OSTP turned to the Center for Risk Management at Resources for the Future (RFF) for assistance in formulating methods for broader use of CRA. From that effort, as described in the preface to this book, RFF organized first a preliminary meeting in August 1993 that pointed the way to further exploration into CRA methods. Funding was found to assist in this exploration, resulting in the February 1994 conference and its papers and, finally, this book.

The initial context of the conference for which the chapter material was prepared should be kept in mind when reading them: they were written primarily for a particular audience—federal regulatory agencies—and for a particular purpose—to encourage the use of CRA. However, because these writings represent the state-of-the-art of programmatic CRA, and because the importance and implications of that art reach far beyond the agencies, the contents of this volume will be valuable to a much wider audience.

THE SCOPE OF THIS BOOK

This book is structured as a logical progression, moving from a description of the CRA process to its history, to general principles, and then to the specifics of how to do it. Taken as a whole, the book outlines the evolution of CRA and its surrounding controversy, summarizes lessons learned from past implementation efforts, and suggests new ways of using CRA. While the individual chapter authors are not in total agreement with each other, they do agree to a remarkable degree both about the desirability of doing CRA and about the principles that should guide the conduct of CRAs.

In the book's second chapter, I outline some key questions, the decision points that an agency must face when undertaking a CRA. Most of the questions are applicable to both types of CRA, but the emphasis is on programmatic CRA. I outline the options for answering the questions and in some cases make specific recommendations.

The chapter by Richard Minard presents the history of CRA, particularly within EPA, and details six of the state and local CRA efforts that EPA has sponsored. The essay reveals much about the methodology and the politics of CRA. It shows how a variety of governments have wrestled with the issues described in the first essay. More state-level CRAs have been done than any other type of large-scale CRA, and Minard's essay distills this experience.

Frederick Anderson's chapter describes the background factors that set the context for the congressional attention to risk. He then describes the positions that the principal stakeholders—environmentalists, industry, state and local governments, and the Clinton administration—have taken on risk assessment and CRA. Finally, Anderson draws a series of lessons and recommendations aimed at improving the use of CRA in the executive branch. He outlines how the administration might promote receptivity to risk-based priority setting while responding to criticisms and concerns.

The chapter by John Graham and James Hammitt makes a series of suggestions to refine the framework for conducting CRAs in the federal government. As they state, "This chapter is intended to provide some conceptual guidance to administration and congressional officials who are engaged in the task of promoting a new priority-setting process informed by comparative risk." Graham and Hammitt particularly stress the importance of ranking risk-reduction options as well as baseline risks—in other words, of looking at the incremental risks that would be averted by specific actions.

The final chapter, written by a group from Carnegie Mellon University (M. Granger Morgan, Baruch Fischhoff, Lester Lave, and Paul Fischbeck) proposes a procedure that agencies can adopt to rank systematically the risks for which they are responsible. Whereas the other authors suggest principles, caveats, and questions, the Carnegie Mellon team says in effect: "If you want to do a CRA, here's how." The methodology it describes is comprehensive and incorporates a number of methodological innovations, such as the use of public panels. The RFF-OSTP workshop spent more time on the Carnegie Mellon paper than on any other because it most clearly spoke to the question of what specific steps an agency would take to do a CRA.

A BASIC VOCABULARY OF RISK

A multiauthored volume about any discipline undergoing transformation inevitably raises problems of terminology. CRA is just such an energetically developing field, so this book is no exception. For quite legitimate reasons, the precise definition and usage of some terms vary considerably among risk professionals, so little effort was made to impose a uniform set of definitions on the chapter authors. Therefore, the same term might have a slightly different meaning and, conversely, the same phenomenon might be referred to by different names. The following paragraphs identify and discuss the basic risk terms that most frequently may cause confusion. These meanings derive from discussions with Paul Portney, and are based in turn on our understanding of the common usage of these terms and their logical interrelations.

Basic Risk and the Four Pillars of Risk Analysis

To start, *risk* itself is simply the likelihood that injury or damage is or can be caused by a substance, technology, or activity. Often a risk estimate contains an explicit probability factor: "There is a 50% chance that the structure will collapse and cause the death of three people."

In similar fashion, when describing the risk-based disciplines, *risk analysis* is the most general term, which encompasses comparative risk analysis, risk assessment, risk management, and risk communication.

Comparative Risk Analysis (Risk Ranking). Comparative risk analysis—CRA—is also termed *risk ranking* or *relative risk ranking*. One type of CRA consists of comparing two relatively well-defined types of risk. It could consist of comparing risks from two similar sources—the cancer risk from two different pesticides—or it could compare two dissimilar risks—the risk of dying in a canoeing accident with the risk of dying from being exposed to benzene. A CRA could also be a comparison of the risk from two different types of controls on the same source, such as comparing the use of chlorine to ozone for purifying water.

A second type—*programmatic CRA*, the focus of this volume—is used for setting regulatory and/or budgetary priorities and involves comparison among a large number of risks. This larger-scale CRA also differs from the first type by involving a great deal more value judgment, and is as much a philosophical as a scientific effort. Examples of programmatic CRA would be establishing the relative importance of various pollution control programs on the basis of risk reductions or establishing priorities based on risk for cleanup among a group of hazardous waste sites.

Risk Assessment. *Risk assessment* is a set of analytical techniques for answering the question: How much damage or injury can be expected as a result of some event? Although risk assessment was first developed to estimate the probabilities of an accident in a particular type of technology (a nuclear reactor, for example), risk assessment methodology has been most elaborately developed for estimating the cancer effects from chemicals based on laboratory testing of rodents. The general term "risk assessment" is often mistakenly equated with this particular type of cancer risk assessment.

The most generally accepted formulation of risk assessment is the following four-step process, devised by a committee of the National Academy of Sciences (NAS 1983).

1. *Hazard identification,* which identifies the type of injury that can be caused (for instance, chemical X can cause liver damage).
2. *Dose-response assessment,* which estimates the relationship between exposure to a harmful substance (or event) and the resultant harm (for instance, exposure of X parts per million of substance Y for a period of 2 hours can produce liver damage).
3. *Exposure assessment,* which estimates how much of a substance will reach a target population or how much of a population will receive some exposure to a substance.
4. *Risk characterization,* which combines the information from steps 2 and 3 to estimate the amount of injury or damage that will be caused by a substance, technology, or other risk source.

Risk assessments can be done for acute health effects such as workplace injuries, chronic health effects such as cancer or birth defects, and ecological effects such as reductions in species or damage to trees. All risk assessments suffer from a fairly high degree of uncertainty. The statistically calculated uncertainty associated with most chronic health risk assessments is several orders of magnitude (there is a hundredfold or thousandfold spread between the high and low plausible estimates). The real uncertainty is even larger because of controversy about the basic premises of most risk assessments, such as the validity of animal testing as a way of predicting human health risks.

Risk Management. *Risk management* developed as a contrasting term to "risk assessment." Risk management considers the social, economic, and political factors involved in risk analysis, determining both the acceptability of damage that could result from an event or exposure and what, if any, action should be taken with regard to the risk of that damage. Compare this to risk assessment, which helps estimate the likelihood that such-and-such damage could result from an event or exposure.

In some instances, risk management refers to all risk-related policy processes other than risk assessment—agenda setting, decisionmaking, implementation, evaluation—whereas in other situations this term is limited to decisionmaking about risk.

Risk Communication. *Risk communication* is just that: conveying information about risk. Such communication can range from simple warning labels to product data sheets to hazardous site databases to public hearings. Originally, risk communication was generally one-way, with the risk experts (whether technical advisor or risk manager) trying to impart expert knowledge to the public. The view that the process needs to be considered as a mutually informing interchange among the interested parties—including the public—is becoming more widely accepted.

A STRONG FOUNDATION OR A WEAK REED?

Congress' consideration of regulatory reform legislation, the effort by state and local governments to sort out sensible federal mandates from arbitrary ones, and the realization by industry, the press, and the public that a process is needed for establishing environmental priorities all lead to a focus on comparative risk analysis. In the short period since the publication of *Unfinished Business*, CRA has gone from being a relatively unknown technique to being at the center of environmental policy and political debates.

The authors of the essays in this book generally share a belief that CRA provides a good conceptual foundation for setting environmental priorities. However, they also are acutely aware of the shortcomings of the approach and of the need to consider a variety of factors other than risk. This book can serve as a guide to those who want to know what CRA can and cannot do. CRA is both a strong foundation and a weak reed, and the book delineates both its strengths and its weaknesses. By providing such a guide, we hope to encourage the application of CRA in ways and circumstances that are appropriate. We also hope to provide a warning to those who would rely on CRA for more answers than it can provide.

REFERENCES

NAS (National Academy of Sciences). Committee on the Institutional Means for Assessment of Risks to Public Health. 1983. *Risk Assessment in the Federal Government: Managing the Process.* Washington, D.C.: National Academy Press.

2

Ranking Risks:
Some Key Choices

J. Clarence Davies

Comparative risk analysis (CRA) of entire agencies or programs—an endeavor also known as *programmatic CRA*—may be done for several different reasons, as noted below in the first of fifteen questions posed to guide the undertaking of such analysis. Regardless of the purpose of a CRA, though, its practice entails a large number of difficult choices. There are no "cookbook" methods for doing such analysis. Large-scale CRA is a relatively new type of undertaking, it is very complex, and it is as much art and value judgment as objective science. For all this—indeed, perhaps because of this—it can also be very useful.

This chapter outlines some of the key choices that must be faced by any agency undertaking a large-scale CRA, identifying the critical questions that must be posed when developing a CRA, outlining the options for answering the questions, and in some cases recommending which option should be chosen. The questions have been grouped into four sections: scope and direction of the CRA; estimating risks; comparing risks; and followup. The questions and the options should be read together because often the best way of delineating the question is to describe the options among which the CRA analyst must choose.

The questions posed in this chapter are, I believe, the most important issues that arise in doing large-scale CRA. However, many other issues could have been raised. The chapter can be read both as an initial agenda of questions for any organization or individual contemplating doing a CRA as well as an outline of some of the limitations of current CRA methodology.

The number of difficult choices to be made and the lack of good options for some of these choices might discourage people from doing CRAs. My intent in this chapter is not to discourage the conduct of CRAs, but to encourage people to anticipate the attendant problems and choices realistically. CRA is a vitally important tool for making health and safety efforts more effective and more efficient. But unrealistic expectations will lead to CRA's being discredited and to CRA efforts' being abandoned. Only with an understanding of the choices and limitations of CRA can we broaden its application and improve the methodology for doing it.

SCOPE AND DIRECTION OF THE CRA

These first five questions are guidelines for the overall scope and direction of the CRA.

- What is the reason for doing the CRA?
- What process is to be used to do the CRA?
- What is to be compared?
- What will the final product be?
- How should noncontrol programs be incorporated into the CRA?

Purpose of a CRA

What is the reason for doing the CRA? Many decisions about the CRA will be guided (or at least should be guided) by the purposes the CRA is to serve. To what uses is the CRA to be put? Who will use it? The answers will determine the process for doing the analysis, the amount of resources devoted to the CRA, and the nature of the final product.

Options. CRAs can be used to allocate budgetary and other resources, to identify problems that require increased attention, and to assist in an agency's obtaining agreement on and support for an environmental agenda. The process of doing a CRA is likely to open up discussion of a broad range of related questions, such as the agency's relationship with the legislature and with the public, the goals that the agency is intended to serve, and how to measure whether the agency's programs are succeeding. Facing such questions is difficult, but raising them may be among the most useful purposes served by CRA.

Some agencies (such as the Department of Defense and the Department of Energy) may want to use CRA to set priorities for cleanups among individual sites. This is a more specific use of CRA than the U.S. Environmental Protection Agency (EPA) report *Unfinished Business* (U.S. EPA 1987) or the state CRA efforts, but most of the choices discussed in this chapter are still applicable. The hazard ranking system used to decide which site to place on the Superfund National Priorities List (NPL) may provide both positive and negative lessons about using

CRA to prioritize site cleanups. On the positive side, the hazard ranking system provides a manageable and useful method for consolidating information about individual sites in a form that lends itself to comparing the relative priority of the sites. On the negative side, the ranking system suffers from a lack of reliable data, the weights it assigns to individual factors have been criticized as arbitrary, and the system's numbers imply a spurious degree of precision. The system is actually designed to provide a yes or no answer to the question of whether the site should be on the NPL rather than serve as the basis for comparing each individual site to all other sites.

One particular issue that has arisen repeatedly is whether CRAs should be used to maximize the benefits obtained for a given dollar amount or should be used to minimize the dollars spent to obtain a given level of benefits. The same CRA can serve both purposes, so this issue does not need to be decided at the outset, but open discussion of it may avoid later misunderstandings, because the political differences between the two positions are significant. Those whose primary goal is to limit government expenditures want to maximize benefits per dollar spent; those who are focused primarily on the benefits to be achieved want to use CRA to achieve the benefits efficiently.

Process Criteria for a CRA

What process is to be used to do the CRA? Estimating and comparing risks are a complex mixture of science and value judgments. Because values play a critical role, *who* does the estimating and comparing is a critical question.

Decisions about the process are essentially decisions about how to define risk, and about what considerations are relevant for establishing priorities. If one believes that priorities should be set only on the basis of relative damage to human health and the environment, then scientists will be the key players. If risk is defined more broadly than damages, then the public has an essential role to play (see the question on other factors for further discussion). However, even under the narrowest definition of risk, risk assessments and comparisons inevitably will involve a number of nonscientific value judgments.

Options. In simplified terms, the key sets of players are: technical experts; agency policymakers; and the public. This is a simplification because the technical experts can be of many different, and often conflicting, kinds (statisticians vs. biological scientists, agency experts vs. industry experts, and so forth); the policymakers can be defined in various ways (by rank, type of responsibility, and so forth; legislators and judges also could be included as policymakers); and there is no single "public." The public includes nonexpert members of environmental organizations and employees of corporations, as well as the nonexperts who have no direct stake in the policy options.

In spite of that simplification, however, these three sets of actors are key because they represent the three basic elements involved in measuring and comparing risk. Each of these three sets of players can be said to have its own expertise in comparing risks.

- The technical experts can provide what information is known about the health or environmental damage caused by a particular environmental insult and how serious the damage is relative to the damage caused by other insults.
- The policymakers bring to the CRA consideration of decision factors such as the cost and feasibility of dealing with various environmental problems, dimensions of equity, and the political viability of various actions.
- The public brings its own set of values, including such elements as voluntariness, dread, and the perceived benefits gained in exchange for environmental damage.

Because of these differing types of expertise, any CRA that does not involve all three sets of players will be seriously incomplete because it will lack one of the basic elements of risk comparison. This does not mean that the only product from the CRA must be the result of the interaction of all players. For example, a technical analysis summarizing what is known about the damage from various types of insults can be a useful product even though later stages of the CRA involve nontechnicians.

Support for the results of the CRA process is another important consideration in deciding who should be involved. Policymakers and the public are more likely to support the decisions that follow from CRA if they have been involved in the process of doing the comparisons.

Types of Comparisons

What is to be compared? A basic question that often is not given adequate consideration at the start of a CRA effort is what types of things are to be compared. The question must be answered at the start of the process and that answer will determine both how the comparisons are made and the uses to which the CRA can be put.

Options. There is a broad range of choices for such comparisons. The categories could include:

- Environmental problems (defined broadly or narrowly, such as air pollution or urban emissions of carbon monoxide from autos)
- Agency programs (again, defined broadly or narrowly)
- Geographical areas
- Specific problem sites
- Proposed actions (such as installation of scrubbers, substitutes for chlorofluorocarbons)
- Economic sectors or sources
- Affected populations

Any of these categories could be used as the basis of CRA. Past efforts have mostly used environmental problems. The choice should be based on the uses to which the CRA will be put. If, for example, the CRA is to be used primarily to establish budgetary priorities, then the categories used in the CRA should be as close as possible to the categories used in the budget. On the other hand, if the CRA is to be used to evaluate environmental justice issues, the categories should probably be affected populations.

Final Product of the CRA

What will the final product be? The more specific the picture of the final product of the CRA—for example, how the rankings will be reported—the easier will be the decision on how to get there.

Options. Once the taxonomy question (What is to be compared?) is answered, the next question is how detailed the comparisons should be in the final product of the CRA. Should the product be a numbered list starting with the highest priority (or riskiest) item and going through to the lowest priority (or least risky)? Or should it be a grouping of items, such as high, medium, and low? If a grouping, how many groups should there be? Again, the answer will depend on the purposes the CRA is to serve. If the purpose is to prioritize site cleanups, a numbered list may be desirable. For most purposes, however, some kind of grouping is adequate.

In most cases a numbered listing implies a degree of precision that is unachievable given the uncertainties inherent in the risk assessments that are one component of a CRA. Assessments of individual risks, for example the cancer risk from a specific chemical, typically have an error range of plus or minus two or three orders of magnitude, that is, the true value could be 100 or 1,000 times greater or smaller than the estimate. Thus, an order ranking based on risk is not likely to be meaningful for individual risks. Comparisons of broad programs or problems (such as indoor air pollution or municipal solid waste) are many times less precise than individual risks because they encompass numerous individual risks.

Programs That Do Not Directly Control Risk

How should noncontrol programs be incorporated into the CRA? Two types of programs do not fit easily into a CRA effort. The first is any programs whose effect on risks is indirect. These include research and development, administration, and policy planning. The second type is any program aimed at preventing new sources of risk rather than dealing with existing risks. The drug approval and food inspection programs of the Food and Drug Administration would be examples. It is, at best, difficult to estimate the magnitude of risks that did not occur because of a preventive program.

Options. I am not aware of any method for incorporating the first type of program in a CRA. In previous CRAs they have not been included in the analysis and, lacking any better alternative, this deletion seems necessary.

Leaving out preventive programs is an option, but not a desirable one because these may be the most important and effective programs in the agency. The risks averted may be estimated from the extent of risk before the program was initiated or by the risk existing in other nations or areas that do not have a comparable program. Some people believe that pollution prevention should be pursued regardless of the amount of risk averted.

ESTIMATING RISKS IN A CRA

These next seven questions are guidelines for the specifics of estimating risks being considered and ultimately ranked by the CRA.
- What types of risk should be considered?
- What time frame should be used for estimating risks?
- How should the existing level of control effort be considered in the CRA?
- Should marginal or total risks be calculated?
- How should uncertainty be factored into the CRA?
- What factors other than health and environmental damage should be included?
- How should transboundary effects be considered?

Types of Risks

What types of risk should be considered? Many different types of risk can be estimated. The choices regarding which ones to include, how they should be grouped together, and other such considerations are critically important for the CRA.

Many proposals for CRA assume that risk can be measured by one endpoint, namely cancer. This assumption arises from the fact that we have made more progress in quantifying cancer risks than other types of risk. To the extent that the assumption has any logical basis, it used to be believed that no adverse efforts occurred at exposure levels lower than the exposures that produced cancer. However, it is now reasonably well established that ecological effects and some serious health effects (such as birth defects or endocrine disruption) can occur at levels below the levels that might produce cancer.

The types of risk considered will determine the results of the CRA. Ranking based on one set of risks (such as chronic health risks) will be completely different from rankings based on a different set of risks (such as adverse effects on human welfare).

Options. The EPA CRAs focused on four risk categories: cancer, noncancer health risks, environmental risks, and welfare effects. With regard to risks covered, on a general level these four categories encompass the major adverse effects of concern. The cancer/noncancer distinction may give too much emphasis to the cancer endpoint. A division into acute health effects and chronic health effects might be more appropriate or, alternatively, all health effects could be grouped together.

Which specific types of risk are included in the analysis are likely to be determined by what data are available. The relationship among which risks are included, how they are measured, and the basis on which they can be compared should be kept in mind (see the questions on risk-ranking methods and comparing different kinds of risks).

Time Frame of Risks

What time frame should be used for estimating risks? If risks are to be compared, the comparison should be for the same period of time. It would not be meaningful to compare the risks controlled (or prevented) by program A over 5 years with the risks controlled by program B over 25 years.

Options. The time frame question is complicated by the difference in latency periods for different diseases. For diseases such as cancer, where 20 to 30 years may elapse between exposure to the causative agent and onset of the disease, a long time frame is necessary. The risk could be counted as controlled at the time of potential exposure rather than at the time of disease manifestation, but this causes more complications, especially if risk is measured in time-sensitive units such as life-years saved.

If risks are measured over a long time period, the question may arise whether risks in the future should be discounted. The answer depends on what use the CRA is to serve, but for most purposes the difficulties of discounting can be avoided by not discounting. One advantage of grouping cancer risk as a separate category is that the importance of the time dimension can be minimized, at least while making comparisons within the category. However, it should be kept in mind that long latency periods are not limited to cancer. Some chronic air pollution effects may take years to become manifest, and genetic effects may take generations.

Current Level of Control

How should the existing level of control effort be considered in the CRA? When comparing the risks of different environmental problems, one needs to ask whether the comparison should be based on the current extent of risk or on what the risk would be if there were no control programs. For example, in the absence of public health programs and facilities, the risks from contaminated drinking water in the United States would be very high. However, the actual risks are low because of existing public health efforts.

Options. As with so many other issues, the answer to this question on levels of control depends on the intended use of the CRA. For example, if the CRA were to be used for incremental budgeting then the existing level of control should be assumed and the relevant question would be the marginal risk reduction that could be achieved for an incremental dollar (see the next question on marginal versus total risks). If, however, the CRA were to be used for zero-based budgeting, then the existing level of control should not be assumed. If the CRA were to be used for prioritizing waste sites for remedial action, then the control question is mostly not relevant, but to the extent that controls exist the extent of risk should be calculated assuming their existence.

Marginal versus Total Risks

Should marginal or total risks be calculated? This is just a broader version of the preceding question on levels of control, but it is sufficiently important to deserve separate consideration.

- Total risks are *all* the risks posed by a problem or program, such as all risks from air pollution.
- Marginal risks are the portion of risks that would be addressed by some particular action, such as the type and amount of risk that would be avoided by removing lead from gasoline or the additional stream miles cleaned up by adding a million dollars to the water pollution control budget.

Frequently, marginal decisions (such as adding dollars to a program's budget) are justified on the basis of total risk (such as the total risks that the program deals with are greater than the total risks covered by competing programs). But it is not logical, for example, to recommend adding funds to the air pollution budget because the overall risks from air pollution are greater than the overall risks from other problem areas. The marginal additional dollars put into air pollution might buy more risk reduction in some other problem area, even though the total risk from the other problem area was lower.

Options. In most (but not all) cases, the decisions for which the CRA is relevant are decisions that involve marginal risks, that is, they do not involve entirely new programs or dealing with problems that have heretofore escaped attention. However, it is often easier to calculate total risks than marginal risks. It is not possible to use a CRA based on total risks to calculate the incremental risks that would be addressed by a specific action. Information on the incremental risks relevant to the decision must be collected and analyzed separately.

Factoring in Uncertainty (Quantitative Risk Estimates)

How should uncertainty be factored into the CRA? Every step of the CRA process involves large degrees of uncertainty. The typical health risk assessment is accurate only within two or three orders of magnitude, many kinds of risks can't be quantified at all, and comparing different kinds of risk introduces uncertainties of huge but usually unspecifiable amounts. Any CRA results that do not make this uncertainty clear are misleading.

Options. The process for doing the CRA and presenting the results should make clear that the conclusions are opinions heavily influenced by personal and societal values and only partially grounded in scientific evidence. Even when this is clear, however, the initial stages of the CRA effort may involve use of quantitative risk estimates to provide some basis or guidance for estimating the overall risk. Such quantitative estimates should be accompanied by estimates of their uncertainty.

A variety of statistical methods exists for calculating and displaying the uncertainty. No single statistical measure of uncertainty is adequate to reliably convey the uncertainty of risk estimates, and thus several different measures should be used (NRC 1994).

There has been some controversy with regard to whether quantitative cancer risk assessments should be based on the "worst case" or should provide the "best" or "most plausible" estimate (NRC 1994, especially Appendix N). This is a less pressing issue for CRA because the comparison will presumably be between risk assessments based on the same assumptions (worst-case figures compared with worst-case figures) and also because the very approximate nature of risk estimates for large groups of problems makes specific quantitative risk estimates less important. However, there is no question that the assumptions built into any risk assessment are critically important in determining the outcome of the assessment.

Other Factors: Psychological, Political, Social, Economic

What factors other than health and environmental damage should be included? This is a critical question, one that is intimately related to the type of process used to do the CRA (see the earlier question concerning CRA process criteria).

Options. Two categories of factors need to be considered. The first are those elements of risk that are used by the public as integral elements in comparing risks. Paul Slovic, Baruch Fischhoff, and others have documented these elements (Krimsky and Golding 1992). They include dread, voluntariness, equity, and so forth. The second category consists of factors that are more clearly not part of risk but which may be important considerations in the decision for which the CRA is being used. This second category includes cost, administrative feasibility, political viability, and so forth.

The first category should be seriously considered as an integral part of doing the CRA. If the CRA is based only on the amount of health and environmental damage, it may be of only limited utility, and may lack perceived legitimacy. Some of the elements of the public perception of risk represent basic societal values that are widely shared and that influence most people's estimation of how important a risk is. For example, society values "real people"—individuals with names and faces—more than the abstraction of statistical individuals. Thus, although kidney disease prevention programs might save many more lives than the same amount of money invested in dialysis machines to keep kidney patients alive, there might be broad agreement to spend the money on dialysis machines. It can be argued that such agreement is not irrational because of the higher value society places on saving identifiable individuals. Risk perception also can directly influence actual damage from risks. For example, the perception that a hazardous waste site poses risk will drive down property values in the vicinity of the waste site. The property values are driven by the perception, not by the actual amount of damage to health.

Risks as perceived by the public are best incorporated by involving the public in the CRA. This can be done by using advisory groups, focus groups, random samples, or other mechanisms.

Equity, in the sense of the fairness with which costs and benefits are distributed, is one of the public factors that importantly influence perception of risk. Recently, environmental equity has taken on the more specific meaning of the effect of environmental problems and decisions on low-income minority communities.

Whether to give special consideration to low-income minority communities is a politically sensitive question. It has been argued that poverty and discrimination against minorities should be considered environmental problems and that low-income minority communities by definition face higher risks than other communities, and thus should be given special consideration in a CRA. How this would be done is not clear, but in any case stretching "environmental risk" to cover all of society's problems may well lead to the term's losing all meaning.

Some environmental justice advocates argue that current methods of measuring risk are misleading and prejudicial because *cumulative* risks that affect the same people or community are not calculated and only population risk, not individual risk, is measured. These are important and sometimes valid criticisms.

The second category of factors (cost, political viability, and so forth) should not be combined with the CRA because the factors are not part of the definition of risk. However, matters such as cost must be kept in mind because they often need to be considered as part of a *decision* along with the CRA results. For example, decisions on site remediation should consider the cost of remedial measures and their technical feasibility, not just the amount of risk that could be eliminated.

Transboundary Effects (Risk Geography)

How should transboundary effects be considered? Almost all CRAs are geographically bounded. In many cases there may be a disparity between the sources of the risks and the places where the risks occur (for instance, pollutants may arise in one state or nation but create risk in a different state or nation). This disparity is potentially a problem if the CRA is linked to programmatic actions because some part of the risk will not be within the control of the entity that can take the actions.

Options. All risks of any given type should be included in a CRA regardless of the source of the risk. However, to the extent that the data allow the analysis to identify the portion of the risk arising from transboundary sources, this adds to the value of the CRA, because it identifies problems where remedies may not be available to the entity doing the CRA.

COMPARING RISKS IN A CRA

These next two questions deal with the specifics of how the risks are considered in a CRA.
• What methods should be used for ranking risks?
• How can different types of risk be compared?

Risk-Ranking Methods

What methods should be used for ranking risks? The methods used for ranking risks will have been largely determined by the choices outlined above in the sets of questions concerning the scope and direction of the CRA and estimating risks. Who will be responsible for the final ranking choices should be decided at the beginning of the process.

Options. The ranking process can be pictured as a continuum ranging from a narrow process heavily determined by scientific estimates of adverse effects to a broad process that incorporates multiple values and is clearly subjective/political in nature. A CRA can fail by going too far in either direction. Technical analyses of adverse effects should be a very important element in a CRA, but if that is the only element, the CRA will be criticized for its failure to include other factors and considerations. If the CRA incorporates all relevant factors, including political factors, and gives them the same weight that they are given in the decisionmaking arena, it will wind up with the *status quo* and will have been unnecessary. If a CRA is a mirror of the more comprehensively inclusive "real-world" decision process, then the analysis adds nothing to the process.

Comparing Different Types of Risks

How can different types of risk be compared? A typical CRA will entail trying to compare very diverse types of risks. However, no scientific methodologies exist to make any of these comparisons. There is no scientific way to compare, for example, health risks with damage to buildings. Values are an inevitable part of the comparison.

Options. Several methods can be used to make such comparisons:
- use dollars as a common metric by estimating the dollar value of each risk (or of avoiding the risk);
- use some common metric other than dollars (this has occasionally been attempted but not with any success); or
- rank the different risks according to some common dimensions, such as severity, duration, irreversibility, geographic scope, number of people affected, and so forth.

A fourth option, the one most commonly used, is to avoid comparisons across unlike categories and compare risks only within the health, environmental, and welfare categories. This may be the best option, but it has three disadvantages. It provides no overall ranking and thus may not serve the purpose for which the CRA is intended. Also, it may be difficult to separate different types of risks. (For example, the problem of stratospheric ozone depletion simultaneously causes acute and chronic health effects, environmental effects, and welfare effects.) Finally, within a limited number of categories it may still be difficult to compare different types of risk. (How, for instance, does one compare loss of a wetland to an oil spill, or a case of asthma to one of lung cancer?)

FOLLOWUP ACTIONS TO A CRA

What followup actions will be taken when the CRA is completed? The answer to this will obviously depend on the reason for doing the CRA in the first place and on what has been learned in the process of doing it. Consideration of followup actions should occur before the CRA is started and should be reconsidered throughout the process of doing the CRA.

Lack of followup action probably has been the greatest single problem with CRAs. There are very few examples of CRAs' having changed actual budget priorities or resulted in new or different policies. It can be argued that the *process* of doing a CRA has beneficial effects apart from any changes that result. The process, for example, can help build consensus, improve communication among those involved in deciding or influencing policy, and raise the level of knowledge and understanding of environmental problems. However, the link between the CRA results and other actions needs to be explicit, and often it is the weakest link in the CRA process.

CONCLUSION: CRA AS THE BEST METHOD AVAILABLE

I have not tried to make the arguments for why CRAs are necessary and useful things to do. That belongs elsewhere (see the chapters in this book by Anderson and Minard, as well as Finkel and Golding 1994). Suffice it to say that establishing health and safety priorities is unavoidable and that CRA is the only analytical process we have for setting priorities, at least in the environmental policy area. The challenge is to improve the methods available for comparing risks and to understand the limitations of CRA as well as its promise.

REFERENCES

Finkel, Adam, and Dominic Golding. 1994. *Worst Things First? The Debate Over Risk-Based National Environmental Priorities.* Washington, D.C.: Resources for the Future.

Krimsky, Sheldon, and Dominic Golding. 1992. *Social Theories of Risk.* Westport, Connecticut: Praeger.

NRC (National Research Council). 1994. *Science and Judgment in Risk Asessment.* Washington, D.C.: National Academy Press.

U.S. EPA (Environmental Protection Agency). Office of Policy Analysis. 1987. *Unfinished Business: A Comparative Assessment of Environmental Problems.* Washington, D.C.: U.S. EPA.

3

CRA and the States: History, Politics, and Results

Richard A. Minard Jr.

For roughly six years the lights have been blazing at those laboratories of democracy, the states, as comparative risk practitioners like me have tried to make the abstraction of "comparative risk" a useful tool for democratic institutions. Among the mistakes we've made has been staying too quiet about our successes and near-successes. As a result, the national debate about the methods and values of comparative risk analysis (CRA) has often appeared disconnected from the discipline as it is actually practiced today. To whatever degree comparative risk analysis at the U.S. Environmental Protection Agency (EPA) ever was "hard" and "undemocratic" (Hornstein 1992), antithetical to pollution prevention or the will of the people (Commoner 1992), or a propaganda tool to "raise the threshold for action—with a demand for absolute proof before anything is done" (McCloskey 1994), the states have made it otherwise.

In truth, the states, cities, tribes, and various foreign institutions have made CRA into many things. The discipline is evolving—and sometimes regressing—as different levels of government adapt it to suit their needs. With some exceptions, these adaptations are making the discipline more democratic, more inclusive, more closely tied to locally defined public values, more honest about its own limitations, and, hence, more likely to be productive.

In this chapter, I explore the discipline's limitations as well as its possibilities. (There are far more real-world examples of the former than the latter.) I start with a primer: a description of the form most

Richard A. Minard Jr. is associate director of the Center for Competitive Sustainable Economies at the National Academy of Public Administration in Washington, D.C. The former director of the State of Vermont's comparative risk project, he was the founding director of the Northeast Center for Comparative Risk at Vermont Law School.

comparative risk projects take and a brief discussion of basic terminology. A political and methodological history follows, with the emphasis on EPA's ambivalent promotion of the discipline over the last decade (particularly William Reilly's vigorous efforts in 1990 to frame EPA's mission in terms of risk and risk reduction); the birth and fitful growth of the state comparative risk projects and their relation to EPA's projects both at headquarters and throughout the ten regions; as well as developments at home and abroad since 1990, including the relevance of comparative risk ideas to the environmental justice movement and "environmental indicators." At the core of this chapter are lessons drawn from an analysis (Minard, Jones, and Paterson 1993) of six completed projects—Washington, Colorado, Vermont, Pennsylvania, Louisiana, and Michigan—and from more than twenty other projects now under way in every part of the country.

THE BASICS OF COMPARATIVE RISK PROJECTS

Any comparative risk project attempts to answer two fundamental questions by implementing a CRA: What are the most serious environmental problems here? How can we most effectively address them?

Most state and city officials who initiate a comparative risk project do so with the hope that answering these questions will help them make better decisions about environmental management. They also hope that the *process* of answering the questions will help them build the political momentum they might need to make changes in policies and priorities. Some officials specifically hope to use the results of the projects as tools to reshape their relationships with EPA; most seem to view the projects primarily as ways to reshape their own agencies and their relationships with their staff and the public. Some see the projects as particularly effective ways to bring the public into agency deliberations and decisionmaking (Minard, Jones, and Paterson 1993).

The Process and the Parts

The typical comparative risk project follows six basic steps:
1. define and analyze the risks posed by the environmental problems facing the jurisdiction;
2. rank the risks in order of their severity;
3. select priorities for particular attention; set goals for risk reduction;
4. propose, analyze, and compare strategies to achieve those goals;
5. implement the most promising strategies; and
6. monitor results and adjust policies or budgets accordingly.

Several projects have lumped steps 2 and 3 together. Unfortunately, only a few projects have had much success with steps 4 and 5. State project participants have usually recommended that their sponsors repeat the effort every three to five years to keep priorities on target, but to date no jurisdiction has formally revised its analysis or rankings.

The comparative risk process is part science and part politics: at its best, it puts up-to-date technical information into the hands of decisionmakers—including legislators and the general public—in a way that enables them to make better political or personal decisions.

Comparative risk projects typically share the following characteristics (Minard 1991):

- *Problem list.* Such a list would be the set of environmental problems to be analyzed and compared. Drafters usually pick about two dozen problems that may range from litter to global climate change. One of the strengths of the comparative risk process is that it encourages people to take a much broader look at the environment than they would if they focused only on a single agency's existing programs, as typically happens in the budget process.

- *Criteria for evaluating problems.* A set of analytical criteria define what the participants think is important to measure, such as pollution levels or various types of risks to human health, to ecosystems, or to a population's quality of life. A list of criteria often will specify what type of units analysts should use for measuring impacts under each criterion (that is, lives lost, dollars lost, rate of change, recovery time, and so forth). Some of the criteria will permit quantitative estimates of harm or risk, but others will require qualitative descriptions of such impacts as aesthetic degradation or injustice.

- *Ranking.* This is the process that participants use to sort out the data and draw conclusions about the relative severity of the problems or their subcomponents. The ranking inevitably involves comparing problems along several dimensions or criteria at once. The result may be an ordered list, a table of scores, a set of graphs, or a concise verbal discussion. Depending on the type of criteria used in the analysis, the ranking may simply relate the relative seriousness of the problems, or it may suggest which problems should be considered higher or lower priorities for action.

- *Action plan.* Most of the projects have attempted to use the information gleaned from the first phase of the process to produce legislation, a set of specific recommendations for new programs, or adjustments to old programs and budget priorities. The most effective of these have identified priorities by comparing the risk reduction potential of a number of alternative proposals. Thus, many of the same analytical tools that the projects develop in the first phase can be employed in the second.

Because of their breadth, CRAs are crude tools. Even when focusing on only a single city, analysts can't use the kind of detailed risk assessments that EPA and others use on site-specific evaluations. They can't measure or even estimate every individual's exposure or response to even a single pollutant. Instead, they have to make sweeping generalizations about pollution levels and exposures, as well as about how people or ecosystems respond to those exposures. In this respect, the projects are like laws or regulations: as the size of the jurisdiction decreases, they tend to produce a better fit.

The projects are large undertakings, requiring experts from many technical disciplines just to answer the technical questions. "Nonexperts" from diverse backgrounds are often required to provide legitimacy to the process of answering the value-laden questions. Most projects have used technical work groups to analyze the problem list, and either senior public managers or some kind of public advisory committee to rank the problems and make other policy decisions. Most state projects have taken at least a year to get authorized and organized, roughly six to ten months to analyze and rank the problems, and as long as a year to deliver an action agenda to a governor, legislature, or city council (Minard, Jones, and Paterson 1993).

Comparative Terminology

As the history section below will demonstrate, the evolution of the comparative risk process has primarily been one of ever-expanding scope. A decade ago, the term "risk" in this context referred principally to calculations of an individual's probability of developing cancer in response to exposure to a chemical or a type of radiation. Although these chemical-specific risk assessments are part of most comparative risk projects, most such projects also consider other health effects that aren't calculated probabilistically, as well as impacts on ecosystems, social institutions, and the economy. Practitioners tend to use "risk" to mean any type of harm or potential harm to things people value. To emphasize this breadth, many practitioners refer to comparative risk *analysis* or a comparative risk *process*, rather than to comparative risk *assessment*. Some prefer the terms *relative risk* or *relative hazard*. Recently, some practitioners have begun to use "relative risk" in a new way, referring to comparisons of problems along a single dimension, such as their impact on human health.

The vocabulary of the field is taking on a charge sparked from the broader political context. In 1985, the administrator of EPA could write that risk assessment "is a kind of pretense; to avoid the paralysis of protective action that would result from waiting for 'definitive' data, we assume that we have greater knowledge than scientists actually possess and make decisions based on those assumptions" (Ruckelshaus 1985). In 1994, the president of the Sierra Club could assert that EPA and practitioners use the word risk to "divert attention" away from the severity of the nation's environmental problems:

> Use of this terminology is part of a broader effort to obscure the role of judgment and values in pollution control and make it sound like it can be addressed solely in terms of a "scientific process." What is not admitted is how much of this "scientific process" really is window dressing. However, the terminology both sounds convincing and acts to exclude the interested public from participating in the process. As such, it also disempowers the public. This emphasis on "scientific process" also is designed to raise the threshold for action—with a demand for absolute proof before anything is done. (McCloskey 1994)

The sharp contrast between these two views suggests to me nothing *inherent* in comparative risk analysis that is empowering or disempowering, mobilizing or paralyzing, proenvironment or antienvironment. Rather, the tool seems to take on those attributes from its users.

THE EVOLUTION OF COMPARATIVE RISK PROJECTS AT EPA

Analysts have been comparing activities on the basis of risks—particularly the risk of premature death—for decades, and for most of those years they have tried to persuade the government to use the comparisons as the basis for setting priorities (Covello 1991). In a political system that is more responsive to the public than to experts, however, priorities tend to follow the public's understanding of problems. Such was the case at the birth of EPA and its statutory framework.

In the early 1970s, the most serious environmental problems were obvious to all: "damages from gross air and water pollution," according to Al Alm, a former deputy administrator of EPA. The new agency and Congress responded with "a holy war against pollution of all forms" (Alm 1991).

The righteous environmental programs the nation adopted in the 1970s collided with President Ronald Reagan's domestic agenda in 1981. Into the rubble at EPA in 1983 stepped William D. Ruckelshaus, whose mission for his second term as administrator included restoring some public confidence in the agency. He sought to do so by strengthening the quality of the science the agency used to justify its actions and by making the science more accessible to the public. In a 1983 article, he observed:

> To effectively manage the risk, we must seek new ways to involve the public in the decision-making process. Whether we believe in participatory democracy or not, it is a part of our social regulatory fabric. Rather than praise or lament it, we should seek more imaginative ways to involve the various segments of the public affected by the substance at issue. (Ruckelshaus 1983)

Two years later, he framed the priorities question in terms that echo throughout the literature today. After demonstrating that EPA's statutes contained goals and deadlines the agency could not possibly meet, he argued:

> We should learn to look at all the impacts of pollution from the perspective of balancing some definable improvement against our always finite resources, whether control expenditures or governmental attention. This is always a hard case to make, especially to Congress, where the prizes seem to go for new programs and ever more stringent ones. It sometimes seems that those who write our protective statutes would rather have EPA pretend to be doing a thousand tasks than have it select the hundred most important tasks and do them well. (Ruckelshaus 1985)

Although the rewards in the 104th Congress now seem to go to those who undo the most stringent programs, EPA's leaders have continued to use Ruckelshaus' vision of analytically sound priorities as a tool to moderate what they perceive as the excesses of Congress.

Unfinished Business: The First Sketch

In 1986, as if responding directly to his predecessor's challenge, EPA Administrator Lee M. Thomas asked seventy-five EPA career staff people to conduct the agency's first comparative risk project: an analysis of the relative risks posed by the thirty-one pollution sources over which EPA had jurisdiction, including, for example, stationary sources of criteria air pollutants, municipal sewage treatment plant discharges, abandoned hazardous waste sites, and leaking underground storage tanks. EPA published the project's results in February 1987 as a multivolume report, *Unfinished Business: A Comparative Assessment of Environmental Problems*. Thomas reiterated some of Ruckelshaus' points in his preface to the report:

Their report—although subjective and based on imperfect data—represents a credible first step toward a promising method of analyzing, developing, and implementing environmental policy. That is why I am presenting it to the public as I have received it.... In a world of limited resources, it may be wise to give priority attention to those pollutants and problems that pose the greatest risks to our society. That is the measure this study begins to apply. It represents, in my view, the first few sketchy lines of what might become the future picture of environmental protection in America. (U.S. EPA 1987)

The project's technical work groups defined four types of risk: cancer risk, noncancer human health effects, ecological effects, and welfare effects. Using the basic risk assessment and economic tools at EPA's disposal, the work groups attempted to characterize the risks posed nationwide by each of the thirty-one pollution problems. The health effects teams estimated risks to the most exposed individuals and to the population as a whole. The ecological risk analysts discussed the problems' impacts on different types of ecosystems. The welfare team tried to estimate the monetary costs associated with the most obvious damage caused by the problems: crop losses, materials damage, and health care costs.

The introduction to *Unfinished Business* acknowledges that the welfare analysis omitted "intangible characteristics that people often find just as important," such as equity, degree of voluntariness, and the value people place on preserving a resource even if they never plan to use it (U.S. EPA 1987).

After estimating the risks, typically with heavy reliance on the analysts' "best professional judgment" to bridge the data gaps, the technical teams ranked the problems in order of the seriousness of the risks they posed. Despite the degree of subjectivity involved in these rankings, the work groups declined to integrate the four rankings into a single list, apparently because the exercise would have clearly crossed over into the realm of value-laden policymaking.

Lee Thomas fully acknowledged the role of values in the rankings when he wrote that the report was "subjective and based on imperfect data," and he seemed to have been apologizing for the subjectivity. He doesn't anymore. In an interview in early 1994, Thomas discussed comparative risk, *Unfinished Business,* and his introduction to the volume: "I think subjectivity is a plus," he said (Thomas 1994).

One of the real problems or mistakes EPA has made, and I was part of it, is taking a concept like risk assessment and using it like it was a scalpel when it's about as accurate as a meat ax. We take a concept that uses numbers, that's what risk assessment is all about, and we turn over decisionmaking to the statisticians, as opposed to taking a concept that is built on a lot of uncertainty, a lot of assumptions, and taking a full picture into a policymaker who can use that as well as a variety of other factors to then make a decision about what's safe and what's not safe.

To me comparative risk needs to be a policymaking tool, not a technocratic, fine-tuned number-crunching tool. [We] need to set some priorities, and we need to use a variety of methods to set those priorities.... I'd like to know what the risks are, and I don't need a statistician to tell me it's ten to the minus four, six, or eight, I need a good discussion of what does the science tell me.... [I need to know] what is the exposure, what part of the population is going to be affected by this, what's going to be the impact on the ecosystem? Is this something that's only going to impact only young children, or is this going to have a long impact that I don't know what it is, because it's starting way up here on the biota, or something. Give me a full picture and let's put down those factors, then I'll make a subjective set of judgments on how I think we ought to order those priorities. (Thomas 1994)

EPA still occasionally seems uncomfortable with the subjective, value-laden component of comparative risk projects (Lash 1994; Minard 1995). This longing to keep the work technical gives Sierra Club president Mike McCloskey and other critics a legitimate reason to be suspicious.

The *Unfinished Business* teams concluded that the problems posing the biggest risks to the nation (such as indoor air pollution and radon, global warming, ozone depletion) were generally not EPA budget priorities. Some weren't even explicitly part of EPA's statutory mandates. *Unfinished Business* is still talked about because it concluded what most subsequent comparative risk projects have concluded: the biggest remaining environmental risks tended to rank low in the public's ranking of risk, as revealed by opinion polls. Problems that the analysts ranked as of low to medium risk (such as those addressing hazardous waste facilities, abandoned hazardous waste sites, underground storage tanks, and so forth) turned out to be among EPA's most expensive programs and among the problems the public ranked the highest (U.S. EPA 1987).

Reducing Risk: Taking Care of *Unfinished Business*

Although *Unfinished Business* failed to change the public's perception of risks and had little immediate impact on most of EPA's programs, senior managers found the project compelling enough to encourage its replication out of town. By 1988, comparative risk projects were under way in three EPA regional offices—Region 1, New England; Region 3, the Mid-Atlantic states; and Region 10, the Pacific Northwest—and in three states—Washington, Colorado, and Pennsylvania. As that work slowly began to snowball, William Reilly took over at EPA and gave CRA a very big push. He asked EPA's independent panel of experts, the Science Advisory Board (SAB), to peer review *Unfinished Business* and offer suggestions on how to respond to any issues it might raise.

The result of the SAB's Relative Risk Reduction Strategies Committee (RRRSC) was a four-volume report called *Reducing Risk: Setting Priorities and Strategies for Environmental Protection*. The SAB acknowledged many of the problems with the comparative risk method, criticized *Unfinished Business* and EPA, but soundly endorsed CRA as a valuable guide for setting priorities. The SAB encouraged EPA to use the method.

Reilly enthusiastically promoted the report at several attention-grabbing events. During a hearing he had requested before the Senate Environment and Public Works Committee in January 1991, Reilly endorsed the report's call for risk-based priorities and argued for significant change at the agency based on the divergence between the SAB's "expert" ranking of environmental risks and the public's judgment, as revealed in a Roper opinion poll (Reilly 1991; Stevens 1991). The SAB had reached conclusions similar to those in *Unfinished Business*: the highest risks to ecosystems included habitat alteration and the loss of biodiversity, global climate change, and ozone depletion. To the extent that the project participants had ranked health risks at all, they concluded that the highest risks included air pollution, indoor air pollution, drinking water contamination, and worker exposure. The public still put hazardous waste sites at the top.

The SAB concluded that "EPA should target its environmental protection efforts on the basis of opportunities for the greatest risk reduction" (U.S. EPA 1990b). Five years later, the National Academy of Public Administration would make a similar recommendation (NAPA 1995) after concluding that EPA's individual program offices had made progress in setting priorities based on risk-reduction potential, but that the agency as a whole had not yet fully realized a risk-based budget. The academy noted that EPA's priorities remain constrained by congressional will and public attitudes.

The SAB had made a similar point when it asserted that EPA had obligations to democratic governance as well. The board emphasized that no amount of science could—or should—eliminate the role of normative values in setting priorities:

> ...the SAB recognizes that risk analyses always will be imperfect tools. No matter how much the data and methodologies are improved, EPA's decisions to direct specific actions at specific risks will entail a large measure of subjective judgment. Yet the SAB believes that relative risk data and risk assessment techniques should inform that judgment as much as possible. (U.S. EPA 1990b)

The Human Health Subcommittee had forcefully made a similar point in its appendix to the report:

> To attempt a relative ranking in terms of severity (or significance) of such disparate health outcomes as birth defects in infants compared to paralysis in older persons requires consideration on many dimensions of the values we place on various members of society, families, and the utility of specific physical and mental functions for individuals and society. Such a comparison requires that the impact of each effect be scored for severity, a process necessitating selection of suitable measures and scales of severity, as well as appropriate weighting factors. (U.S. EPA 1990b)

Because of the subjective nature of ranking various types of health effects, the subcommittee recommended that laypeople be involved in the process:

> One possible way to accomplish this is through the use of lay and professional focus groups meeting separately and then together. The process by which this is done, whatever it may be, the way in which the views of informed potential sufferers (and how they become informed) and of medically and technically trained experts are brought together is critical to developing severity factors or indices with any validity or credibility. (U.S. EPA 1990b)

These words were largely ignored at the national level, where the CRA debate remained fixed on the idea of risk analysts' dictating policy. Practioners in the states, however, were more receptive to the advice of the Human Health Subcommittee. By the time *Reducing Risk* was published, the states of Washington and Vermont had already demonstrated the potential for lay citizens and technical experts to work together productively on comparative risk projects. These and other state projects have expanded the definition of CRA. Participants in projects just getting under way in mid-1995 are continuing to search for ways to make their work more effective through increased public involvement, and both EPA and its Science Advisory Board are debating whether and how to undertake a new national comparative risk analysis.

THE STATES AND CITIES: PROCESS, POLITICS, RESULTS*

As the map (Figure 1) illustrates, some three dozen states, cities, tribes, and regions have been or were engaged in some form of comparative risk work sponsored by EPA's Regional and State Planning Branch as of summer 1995. Additional states and cities are actively considering initiating projects. More than a hundred people from around the nation attended the third annual EPA-supported conference for comparative risk practitioners at the end of January 1994. The National Governors' Association called for more risk-based environmental priority setting (NGA 1993, NCCR 1993b). An observer might well ask, Why the groundswell of interest? Is there any indication that going through one of these processes does any good?

The answer is mixed. Comparative risk projects have not radically altered the way states and cities manage the environment or make regulatory decisions. In a few places, governors, legislators, and the public have responded to the projects by endorsing new programs, changing spending priorities, or trying alternatives to command-and-control regulation. These changes will be examined below. In no place has a comparative risk project spawned a technocratic threat to ordinary democratic decisionmaking. To my knowledge, there are no teams of scientists and economists huddled in back rooms imposing strictly "rational" policy on the masses. (This widely held fear, too, will be examined below.) Indeed, the states' work with CRA suggests that the most effective projects were also the most democratic.

In state capitals, just as in the nation's, making any fundamental change in policy or approach is enormously difficult in the absence of a crisis. The state and federal bureaucracies that have grown up to implement federal statutes generally resist change. So, too, do the legislative committees that claim jurisdiction over the many fragmented pieces of environmental management. Interest groups and the press have staked out firm positions on most issues. The result is a fairly stable status quo that, as EPA's two national comparative risk projects demonstrated, provides precisely what the public has most vehemently asked for: protection from the risks it most fears or abhors (Breyer 1993).

---

*This section draws heavily on a 1993 study (Minard, Jones, and Paterson 1993). The authors analyzed the six state projects that had been completed as of January 1, 1993, interviewing dozens of project participants and surveying dozens more.

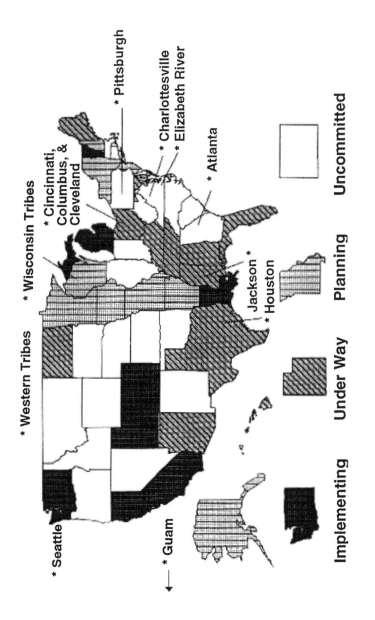

Figure 1. Comparative Risk Projects: May/June 1995

As Ruckelshaus said, whether one believes in participatory democracy or not, public policy ultimately flows from the people. And the only way to change public policy will be with the people's blessing. Thus, it should come as no surprise that the most effective comparative risk projects have been the ones that set out specifically to include key representatives of the public in addition to technical experts.

The projects have an impact on their participants, whoever they are, as long as the individuals take their work seriously. Most have. The challenge of ranking environmental risks demands that participants learn a great deal about a comprehensive list of problems. If the analytical criteria are sufficiently broad, participants almost certainly will have to think critically about the technical and social aspects of those problems in ways they never have before. If the decisionmaking body is sufficiently diverse and has spent any time listening to the public or interest groups, its members will be keenly aware of the role their personal values will play in ranking problems. The ordeal of working as a group to rank problems forces group members to clarify their own thinking as they search for points of agreement with their colleagues or sharpen points of disagreement. The ranking process exposes weak arguments, poor data, and fuzzy thinking. The process tends to break down preconceptions about the problems. The process also breaks down individuals' prejudices about the other participants. The result: members of ranking committees have discovered that they agreed on far more than they had expected. They have come to share a strong conviction that their insights are important and *should be used* to influence public policy. In short, the process has frequently built coalitions for change (Minard, Jones, and Paterson 1993). Experience in the six completed projects has shown that the effectiveness of those coalitions depends on many things, including:

- the membership of the project's committees;
- the commitment and continuity of political leadership; and
- tight connections between the risk-ranking phase and the risk management phase.

I will examine aspects of the first six state projects to point to specific strengths and weaknesses and to illustrate some general points.

Washington Environment 2010

The state of Washington's comparative risk project—Washington Environment 2010—was one of the first three state efforts, and it remains the most successful in terms of its influence on legislation and state policy. Initiated by Christine Gregoire when she was director of the state's

Table 1: Washington 2010 Ranking of Top 23 Threats to the Environment.

Priority Rankings[a]	Environmental Threat
Priority level 1	Ambient air pollution
	Point-source discharges to water
	Nonpoint-source discharges to water
Priority level 2	Drinking water contamination
	Uncontrolled hazardous waste sites
	Wetlands loss/depletion
	Impacts on forest lands
	Impacts on agricultural lands
Priority level 3	Indoor air pollution
	Hydrological disruptions (dams)
	Global warming and ozone depletion
	Regulated hazardous waste sites
	Nonhazardous waste sites (landfills)
	Impacts on recreational lands
	Pesticides (not covered elsewhere)
Priority level 4	Indoor radon
	Radioactive releases
	Acid deposition
	Accidental releases
	Impacts on range lands
Priority level 5	Nonionizing radiation
	Materials storage (tanks)
	Litter

[a] Each priority level signifies a distinctly higher risk management priority than the next level. Threats within each priority level are *not* ranked.

Source: Washington Department of Ecology 1989

Department of Ecology under Governor Booth Gardner in 1988, the project included technical committees drawn from nineteen agencies of state government, a steering committee composed of senior managers or the heads of those agencies, and a public advisory committee composed of thirty-four prominent people from all regions of the state, including legislators, representatives of important interest groups and sectors of the economy, and local and county government officials.

The technical staff researched the problems, then spent many hours briefing the advisory committee, which ultimately ranked the problems in order of their priority. Table 1 shows these rankings, with ambient air pollution and water pollution at the top of the list of threats; hazardous waste problems are in the middle.

After publishing the findings in *The State of the Environment Report*, the advisory committee and state government hosted an open two-day environmental summit in November 1989 to draw the public's attention to the work and to begin soliciting its response. Six hundred people attended. The advisory committee followed up with twelve town meetings across the state that drew another thousand citizens and gave them a chance to learn about and rank the same problems. Meanwhile, a group of analysts, most of whom had been part of the original technical committee, considered the risk reduction potential and costs of some 300 policy options, many of them proposed by citizens at the public meetings. The staff presented their analysis and recommendations to a joint meeting of the steering and advisory committees in early 1990 (WDE 1990b). The decisionmakers drafted an action agenda, took it out for another round of public hearings, and published a final version in July 1990. It listed twelve broad "challenges" for state action and proposed strategies for each. To expedite the release of these challenges and strategies, the committees did not rank or prioritize them in any way.

By the end of the following legislative session, the state had adopted several new environmental laws, dealing with clean air, transportation demand, water conservation, recycling, growth management, and energy policy. The Department of Ecology received appropriations of $16.2 million to support Environment 2010 programs and it also "reprogrammed" $6.8 million from lower to higher risk management priorities.

Gregoire maintains that the project had a profound and positive impact on the Department of Ecology and the state's environment. In a 1992 address to other state practitioners, she said, "Our planning system is undergoing remarkable change as a result of Environment 2010 and is evolving to focus more on strategic (compared to operational) planning with a focus on measurable objectives, performance/results evaluation, and environmental indicators as overall measures of progress" (Minard, Jones, and Paterson 1993). (It is worth noting that Gregoire built upon the positive statewide recognition she gained through the Environment 2010 process to win election as state attorney general in 1992.)

Exactly how much influence did Environment 2010 have on public perceptions and the political climate? Opinions differ. In 1992, the Northeast Center for Comparative Risk (NCCR) at Vermont Law School conducted a survey of participants in state comparative risk projects, including Environment 2010. Several of the respondents suggested that the Department of Ecology has claimed too much credit for itself and Environment 2010. The report of that survey (Minard, Jones, and Paterson 1993) includes the following summary of interviews and correspondence, which begins with a critical comment from Gary Wilburn, counsel to the U.S. Senate Environment and Natural Resources Committee:

"In summary, the Environment 2010 project was only indirectly influential in shaping the environmental focus of the

Washington State Senate. It helped to move the Executive Branch and certain interest groups toward a consensus that major air quality legislation was necessary in light of its high ranking under Environment 2010. However, many would argue that the Department of Ecology had already identified the air quality program as in need of major expansion relative to other programs, and the Environment 2010 findings only confirmed what was already known.... Additionally, in the same legislative session the other major legislation proposed by the Governor was a comprehensive oil spill prevention and response measure. By comparison with the air quality concern, oil spills were not even ranked on the Environment 2010 listing of threats, yet this measure received equal stature in support from the Governor's Office. There is therefore considerable doubt that the Environment 2010 project was the primary mechanism shaping the legislative agenda during the 1991 session."

Bob Nichols, a senior executive policy assistant with responsibilities for environmental issues in the governor's office, criticized the project for being "pseudo-science," for putting far too little emphasis on natural resource issues (more about this later), and, as a result, for failing to develop a truly comprehensive environmental strategy. Nichols, however, gives Environment 2010 credit for making the air legislation part of the governor's priority package and for its passage: "the degree of concern over the air quality problem was a surprise to everyone, and the need for nonregulatory approaches was given a boost."

The Department of Ecology got so hung up on its rankings, Nichols wrote, that it lost sight of more traditional political opportunities. "For example, oil spill prevention did not rank high in the risk exercise and as a result the 2010 proponents felt we should not do anything in particular in this regard—despite the fact that Washington has had enormous problems in this area, the Governor's exposure was hanging out if we had another spill, and the timing was absolutely right to take care of this problem. The 2010 proponents felt they should put full attention on the top priority, and let the lesser priorities fend for themselves." (Minard, Jones, and Paterson 1993)

Nichols reiterates Wilburn's observation that Governor Gardner pushed the oil spill legislation without the help of Environment 2010.

Similarly, Nichols also complains that the project ignored the old-growth forestry issues because they were simply too hot and divisive.

I draw several conclusions from the Washington example:

- The comparative risk process changed what was politically possible by raising public awareness of the risks from air pollution (particularly from automobiles) and exploiting the coalition that formed through the advisory committee.
- The process did not preclude the governor's or legislature's ability to be politically opportunistic. The ranking concentrated the Department of Ecology's attention but did not put anyone in the kind of straitjacket some critics predict.
- The project succeeded in taking more science to more people and thus opened up opportunities for public involvement in the policy debate, rather than the other way around, as critics predict.
- The nontechnical members of the advisory committee demonstrated a solid ability to understand the technical issues and to use that information to make decisions that earned the respect of most of the technical staff and their managers.

The Environment 2010 story ends in a slow fade. The Department of Ecology let the advisory committee drift apart in 1990, failing even to keep them up to speed on developments, let alone exploiting their potential for ongoing support and guidance. The interagency steering committee functioned as a coordinating body for about two years longer. The Department of Ecology's planning staff continued to try to institutionalize the focus on risk by developing environmental indicators to track trends and measure them against the state's goals. As Governor Gardner was preparing to leave office in 1992, he told a reporter that his greatest environmental achievement had been putting in place the long-term risk reduction agenda created by Environment 2010 (NCCR 1992a). After he stepped down in 1993, however, both his replacement and the new director at the Department of Ecology chose to reorganize the department, essentially dismantling the planning staff and state government's memory of Environment 2010.

Analagous experiences abound in the other completed state projects.

Colorado Environment 2000

The Colorado Department of Health's comparative risk project was similar to Washington's in general approach, goals, and timing. It produced far less dramatic results, however, at least as measured in legislation. (A more significant result—an experiment in EPA grant flexibility—is discussed later in this chapter.) A major structural difference between the two projects helps explain the different results.

Colorado's technical work groups analyzed and ranked thirty-one of the state's pollution and resource problems, publishing a summary of their results in 1989, *Environmental Status Report.* The work groups turned this report over to the forty-three-member Citizen Advisory Committee (CAC) whose task was to draft recommendations for action. Governor Roy Romer conducted two public summits with the advisory committee in conjunction with Earth Day 1990.

Only a few of the members of the advisory committee had had any contact with the technical phase of the project, so *Status Report* had no particular meaning for them. The diverse group chose to act by consensus as it developed a list of more than 400 environmental management proposals, opting for inclusiveness rather than decisiveness. Anyone could add to the list as long as no one objected. The committee did not attempt to estimate the proposals' potential costs or benefits, nor did it rank their relative importance.

The 1993 NCCR report on state comparative risk projects discussed these matters (Minard, Jones, and Paterson 1993): given the advisory committee's decision to remain apolitical and above the fray, it is hardly surprising that its members paid little attention to the technical work groups' risk analysis and ranking. Those details would only have been essential if the committee were making hard choices about priorities and how to get the most risk reduction out of each public dollar. According to members of both the technical work groups and the advisory committee, the CRA served only as one piece of information on which the advisory committee based its proposals. The final report of the group includes only occasional, indirect reference to *Environmental Status Report,* and does not include any reference to the risk rankings.

Moreover, the advisory committee disbanded after publication of *Final Report.* The technical work groups had already stopped meeting. No one sought followup funding, and little effort was made to oversee implementation of any of the specific steps included in the proposals.

The Colorado effort seems to have involved all the right people but not in the right way. By skipping the hard work of ranking problems or setting priorities, the advisory committee developed no cohesion or mission. By failing to work with the technical information, the committee developed no particular insights or credibility.

Despite these weaknesses in the project, Tom Looby, the director of the Colorado Department of Health, remained committed to the goals for risk reduction expressed in the two reports and incorporated them in his department's work plans and the state's comprehensive agreement with EPA.

Pennsylvania

Pennsylvania's environmental departments started a comparative risk project at about the same time as Colorado's and Washington's, but they followed the quiet, in-house technical approach used by EPA at the time. The state employees on the technical teams produced a ranking of the state's problems and a report describing their findings, but senior managers who had not been involved in the project decided that the document did not warrant distribution. Apparently, it diverged too far from the administration's policies.

The silver lining of this otherwise gloomy story is the state's demonstration that public managers are not likely to tolerate runaway technocracies.

Vermont

The organizers of the Vermont project designed their committees to withstand a probable change of administrations, including a change of senior public managers. It almost worked.

The Agency of Natural Resources (ANR), working with the departments of Health, Agriculture, and Development and the Governor's Office of Policy Research, initiated the project in 1989. The secretary of the agency, Jonathan Lash, appointed a twenty-member Public Advisory Committee (PAC) to oversee the project from the start. He selected the members to capture Vermont's real diversity and interests, as opposed to picking the leaders of key interest groups. Unlike other projects, the PAC had authority over the problem list and the analytical criteria, as well as the risk ranking. Technical work groups of state employees did the analysis and met several times with the public members to discuss their work and results. The advisory committee members asked for and received briefings by nationally known experts on health risks, ecological risks, and economics.

The advisory committee ranked the state's environmental risks a few days after Vermonters elected a new governor, Richard Snelling. The advisory committee had done a good job, working aggressively to bring the public into the project and gaining insights into the state's problems. The election, however, crowded their work out of the news, and the members' lack of institutional constituencies provided them with no political clout. Thus, it was easy for Snelling and his new head of ANR to ignore the committee members and their recommendations for several months. When the new administration reorganized the risk management phase of the project, it included neither the advisory committee members nor most of the state employees who had worked on the technical committees. These events, plus the upheaval caused by Governor Snelling's death in the summer of 1991, dissipated the project's momentum and muted its impacts.

State Toxicologist Dr. William Bress of the Vermont Department of Health described some of the project's benefits in a note to his colleagues in other states:

> My first impression... of the comparative risk process was that it was a huge waste of time.... I felt that I knew what the comparative risks for different problems were. Why should I spend many hours confirming something I already knew?... My opinion of the process started to change after my first meeting of the citizens' committee. When I saw how they were using the data we supplied, I was pleasantly surprised. Groups of people from all walks of life were often ranking problems in the same order as I had. They came up with these rankings on their own.... Risks we had high on our [the Department of Health's] priority list are now a high priority to the public. This has resulted in shifting of funds and resources to these areas of concern. This might not have happened without going through this useful, although painful, process. (Bress 1992)

As Bress's comments indicate, even when politically diverted, the comparative risk work helped the Department of Health to move more aggressively on lead paint problems and indoor air pollution. (The legislature passed legislation to reduce risks from lead in 1993.) The project also influenced ANR actions, including renegotiating air toxics regulations, preventing staff cuts in the office responsible for protecting rare species and habitats, and setting some crossmedia and interdepartmental work in motion. (For more on the Vermont experience, see the later section on analytical approaches.)

Louisiana

Political upheaval almost did in the Louisiana Environmental Action Plan (LEAP to 2000). When Edwin Edwards replaced Charles E. (Buddy) Roemer as Louisiana's governor in 1992, Edwards' new team fired most of the staff people who had managed the half-completed project. But LEAP's Public Advisory Committee (PAC) refused to go away, and because it was made up of influential people representing major industries, environmental groups, and public interest organizations, the new governor listened. About a year after taking office,

Edwards signed an executive order asking the advisory committee to resume work with his agency people on developing an action plan (NCCR 1993a).

Before LEAP, Louisiana had had no tradition of participatory government and had seen little cooperation between the petrochemical industry and environmentalists. LEAP changed that. The PAC members worked loosely with state agency officials, conducted public forums and a summit together, and made all decisions by consensus. The PAC members not only agreed on a ranking of environmental priorities—which became known as the state's first "environmental treaty"—but believed they could agree on improvements in environmental management.

Members of the PAC have continued to work with the staff of the Department of Environmental Quality. By mid-1995, these efforts had yielded strategy proposals for addressing six of the top ten high-risk priorities and momentum from the project was contributing to a state-sponsored summit meeting on coastal wetlands restoration (NCCR 1995).

Michigan

The most dissimilar of the first six completed state projects was conducted by the Michigan Department of Natural Resources under the strong leadership of William Cooper, an aquatic ecologist at Michigan State University and the chairman of the SAB's Ecology and Welfare Subcommittee. Cooper persuaded distinguished Michigan scientists from academia and industry to serve on the Science Committee for this project. State, county, and local government officials comprised an "Agency" committee. And well-connected people from Michigan's public commissions and various state and private organizations comprised what Cooper called the "citizens with clout" committee. The three groups had equal say in the selection of the problems for study and in the final ranking in March 1992, though the scientists were clearly the first among equals.

The Michigan Relative Risk Assessment Project used the most prestigious group of scientists of any state project up to that point, yet the project was also the least quantitative and the least rigorous. Where most projects produced hundreds of pages of technical analysis drawn from state monitoring data, the Michigan scientists prepared brief white papers on each issue, often with little or no Michigan-specific data. The agency employees reviewed the white papers but had relatively little contact with the scientists. Each of the three committees ranked the problems separately before consolidating their judgments in a final ranking. Afterwards, Cooper was pleased to report that the process had proved his hypothesis that the groups would draw similar conclusions from the information. "They converged like mad," he said (Minard, Jones, and Paterson 1993).

The project was the fastest state effort to date: just ten months from kickoff to ranking. It was also the least outward looking, holding just four public sessions. Cooper designed the process not to reach a broad public audience but to deliver to newly elected Governor John Engler a focused, risk-based environmental agenda that Engler could then work into the regular political process. The governor responded by endorsing the project's report and asking his agencies to develop specific action plans to follow through on its general recommendations. Almost a year of delay followed, apparently because of a lack of support for the project within the bureaucracy. In late spring of 1993, however, the Department of Natural Resources took up the challenge with vigor, appointing task forces to work on each of the five highest-ranked problems.

By mid-1994, the project was regaining momentum under the direction of the Michigan Natural Resources Commission and the Michigan Department of Management and Budget. By the spring of 1995, four task forces had sent reports to the governor, five more were working on reports, and nine more task forces were expected to be appointed.

According to John Shauver, who coordinated the project for the Department of Natural Resources, the task forces were working well because they have had strong participation from academics and interest groups, as well as agency staff. Each of the task forces has included at least one member who served on one of the project's risk ranking committees, so there has been substantial continuity to the ranking process. The task force members discovered that to make thoughtful policy recommendations they first needed to collect and analyze the kind of state-specific environmental data that most projects have gathered *before* their rankings. As a result of this additional information, some of the task force members were changing their minds about the risks associated with the problem areas.

Shauver anticipated that the task force would create strong political coalitions that would encourage the governor and legislature to follow through on task force recommendations. Engler has said he will build his environmental agenda out of these recommendations.

ANALYTICAL APPROACHES: THE ROLE OF SOCIAL VALUES

Variations among the states' approaches to the analytical core of their projects highlight some of the strengths and weakness of CRA. These variations have included different approaches to the problem list, the criteria used to evaluate the problems, and the process used to rank the problems. The variations in the evaluative criteria and in how these criteria influence the rankings reflect the unique social and philosophical values underlying each project.

Most of the projects have focused their analysis on understanding the nature of a given state's environmental problems rather than on the potential effects of risk-reduction options. Thus, the following discussion of the role of social values details the types of choices project participants have made in defining and ranking problems. Most of those choices are value-laden rather than objective. Washington, and to a lesser extent Vermont, attempted to analyze the risk reduction potential of an array of risk management strategies. Analyses of strategic options were particularly useful in Washington. In form and content, the analysis of risk-management options has generally resembled any other evaluation of policy choices, with one twist: because the projects start with a wide range of problems, analysts try to find policies that will address several of the strategies simultaneously, as modeled in *Reducing Risk* by the SAB's Strategic Options Subcommittee (U.S. EPA 1990b).

Problem Lists: Their Scope and Taxonomy

Every project has struggled with the problem list: what to include, how to lump or split the related issues, and what words to use.

Most project lists look something like EPA's thirty-one problems chosen from *Unfinished Business*, though most lists are substantially broader in scope, including resource, habitat, and even social problems that fall well beyond EPA's jurisdiction. Thus, in addition to ambient air pollution, nonpoint-source discharges to water, global climate change, and the like, Washington, for example, considered "litter" and "impacts on range lands." Vermont included the "visual and cultural degradation of Vermont's built and natural landscape." One of the first cities to conduct a comparative risk project, Seattle, considered the "loss and degradation of the urban forest." Michigan included "lack of environmental awareness" and "the absence of land use planning that considers resources and the integrity of ecosystems." A new project in Jackson, Mississippi, includes "poverty" and "AIDS" on its list of environmental problems.

Several projects have looked for the deeper social causes of the "symptoms" of pollution. Thus, Vermont's work groups prepared reports on Vermont's contribution to ecosystem degradation outside the state, the impacts of Vermont's population growth, and the inadequacies of environmental regulations. Because these issues were different in kind from the rest of the problems, the advisory committee chose to discuss them in its report but not to rank them. Arguably, Michigan's "lack of environmental awareness" should have been treated the same way because of its open-endedness. The Michigan committees ranked it among the state's most serious problems, however, and that seems to be helping to motivate the legislature to pass an environmental education bill after twenty years of controversy (Shauver 1994).

As these examples show, the state and city projects generally are avoiding technical jargon in their problem lists and are defining problems specifically to answer their *public's* questions. Often, in fact, the problem titles or definitions are taken directly from requests made at public hearings. Participants are putting a higher premium on using the list as a risk-communication tool than on its intellectual consistency. The SAB subcommittee reports had faulted EPA's problem list in *Unfinished Business* for similar flaws: all of the lists mix sources, receptors, types of pollutants, and social consequences together in a list that makes rigorous comparisons difficult or impossible.

The Criteria: Expanding Definitions of Welfare

Just as state projects have expanded their problem lists to answer public questions, so too have they expanded the analytical criteria by which they measure the relative magnitude of the problems.

Most of the projects have had separate teams to analyze risks to human health; to ecosystems; and to what is variously called social welfare, the quality of life, or simply societal impacts. (Colorado, in contrast, assigned its analysts to teams focusing on specific environmental media: air, water, land, resources.)

The health risk work generally has followed EPA's model, usually employing EPA's cancer potency factors and data on noncancer effects to generate estimates of aggregate population risk and risks to highly exposed individuals or highly sensitive subpopulations. With the growing prominence of the environmental justice movement, states and cities are becoming more careful to discuss the geographical and socioeconomic distribution of risk. Most of the states have tried to estimate or model exposure rates from actual monitoring data. Most have used actual incidence data when available. When states have virtually no data of their own—on some types of indoor air pollution, for example—they have extrapolated from national estimates.

From the beginning, the projects have been clear about the lack of precision in their work, and the newer projects are increasingly blunt about uncertainty. Most are trying to report all estimates in ranges, as ranges, as opposed to points. Unfortunately, none has yet come close to meeting the standards suggested by Adam Finkel (Finkel 1990), who demonstrated a variety of techniques for quantifying and illustrating uncertainty, and thus presenting a more meaningful picture to decisionmakers.

Teams producing the ecological risk analyses have used a variety of approaches to estimate impacts. Washington and Colorado struggled to use the *Unfinished Business* approach. EPA offices in Regions 4 and 6 worked particularly hard to develop an approach that would capture transboundry issues and that could identify particularly vulnerable ecosystems (U.S. EPA 1990a). Vermont used the SAB's approach. Hawaii replaced these broad-brush approaches with some 350 site-specific evaluations. In most cases, the ecosystem analyses have been more impressionistic and qualitative than the health analyses. Despite the simplicity of the tools, analysts were relatively confident that they could make distinctions among certain types of problems, and the results have been fairly consistent nationwide.

Vermont's Values: Quality of Life Criteria

Vermont's approach to the "quality of life" question illustrates how this part of the analysis is changing.

The assignment of project participants was to answer the open-ended question, "What environmental problems pose the most serious risks to Vermont and Vermonters?" The Public Advisory Committee (PAC) began its work by asking Vermonters what they thought the most serious problems were and *why*: What was it about the problem that made it objectionable? The answers came back through eleven public forums, as well as more than 400 responses to a survey designed to elicit Vermonters' values and perceptions.

Vermonters often said that they abhorred water pollution and incinerators, for example, because these two problems threatened their health and the health of Vermont's ecosystems. Citizens of the state also said that they abhorred some problems because of their impact on future generations, or because they were unfairly imposed on people, or because they threatened their property values or their ability to relate to their land the way their families had for generations. Through these answers, Vermonters defined "those other risks" and gave the advisory committee a sense of their relative importance.

The PAC and the project's Quality of Life work group consolidated what they learned about Vermonters' values in a set of seven criteria (see Table 2).

Table 2: Vermont's Quality of Life Criteria.

Criteria	Examples of negative impacts
Aesthetics	Reduced visibility Noise, odors, dust, and other unpleasant sensations (such as water weeds or turbidity in a lake, grime on buildings) Visual impact from degradation of natural or agricultural landscapes
Economic well-being	Higher out-of-pocket expenses to fix, replace, or buy items or services (such as higher waste disposal fees, cost of replacing a well, higher housing costs) Lower income or higher taxes due to the problem Net loss of jobs because of the problem Health care costs and lost productivity
Fairness	Unequal distribution of costs and benefits (costs and benefits may be related to economics, health, aesthetics, or any other of the quality of life criteria)
Future generations	Shifting the costs (such as economic costs, health risks, ecological damage) of today's activities to people not yet able to vote or not yet born
Peace of mind	Feeling threatened by possible hazards in the air or drinking water, or by potentially risky structures or facilities (such as waste sites, power lines, nuclear power plants) Heightened stress caused by urbanization, traffic, and so forth
Recreation	Loss of access to public and private recreation lands Degraded quality of recreation experience (such as spoiled wilderness, fished-out streams, dammed whitewater)
Sense of community	Rapid growth in population or the number of structures in town; development that changes the appearance and feel of a town Loss of mutual respect, cooperation, ability, or willingness to solve problems together Individual liberty exercised at the expense of the common good Community authority exercised at the expense of the individual Loss of the working landscape and the connection between people and the land

Source: Vermont Agency of Natural Resources 1991

Analysts had to find ways to measure, estimate, or describe the risks that each stressor from each problem posed to each of the criteria. In many cases, this required starting with an estimate of the stressor's impact on human health or natural systems, and then estimating how those effects altered Vermonters' quality of life. For example, estimating the economic impacts of a pollutant typically required knowing the biological consequences first. In other cases, such as visibility degradation from air pollution, the quality of life effects could be measured directly. In a few cases, there were useful data in hard units. In most, analysts had to find persuasive qualitative ways to characterize the impacts.

They never did find a good way to capture the "sense of community." Three of the seven criteria offer particularly useful examples of the challenges comparative risk projects face when they depart from the narrower definitions of risk.

Fairness. Vermont's project was the first to make fairness an explicit consideration. Residents had told the PAC that they cared deeply about the distribution of risks and benefits. In an ideal, moral world, there would be no involuntary risk, and the same people who bear the risks from a source of pollution would also reap the benefits derived from it, whether it be a factory, a landfill, or an automobile. When the beneficiaries shift the risks to others, however, the resulting social injustice is an affront to the quality of life. This working definition of fairness captures much of the sense of outrage that people typically feel about "involuntary" risks.

With a little critical thinking, the participants found the fairness criterion remarkably easy to apply. The following summary statements from the advisory committee's report show how the committee distinguished between air pollution and solid waste on the grounds of fairness:

> *Air Pollution:* Because much of Vermont's air pollution is generated outside Vermont in the production of goods and energy not destined for Vermont, the impacts of the pollution are inherently unfair. Vermont's own pollution generators, however, are typically not factories or other large point sources, but the cars, oil burners, and wood stoves used by almost everyone. The risks from these sources, including much of the health risk from air toxics, are more fairly distributed. However, the risks from air pollution, whatever its source, are involuntary.
>
> *Solid Waste:* The advisory committee gave considerable weight to the unfairness of solid waste. Although everyone benefits from having a place to take his or her trash, the landowners adjacent to landfills and incinerators must bear a disproportionate burden of the quality-of-life risks: increased truck traffic, elevated noise levels and other aesthetic affronts, potentially increased health risks, and potentially depressed property values. The fairness equation is reversed for illegal disposal [dumping or backyard burning]: a few people benefit and many others bear the aesthetic and health risks. (VANR 1991)

The advisory committee's comfort with the fairness criterion suggests that other projects might use similar criteria to consider how different problems affect poor communities or communities of color.

Those problems that are more "unjust" might rank higher overall. This kind of analysis could help augment a search for high-risk locations or "hot spots."

Peace of Mind. Scoring problems for their impact on "peace of mind" was as simple as listening to what people said they were afraid of or to what their representatives and lobbyists demanded in the legislature. The criterion tended to capture the outrage factors that underlie the public's risk perception and posed an interesting challenge to the committees when they had to rank the problems. In its final report, the PAC explained its choice:

> Environmental problems, particularly those that may cause cancer, can shatter people's sense of security. The resulting fear can be emotionally debilitating and even financially devastating. Because the Advisory Committee members considered peace of mind an important part of Vermonters' quality of life, they adopted it as a criterion against which to measure the seriousness of environmental problems. The criterion poses a unique problem, however: peace of mind may have little to do with an objective measure of risk or safety. Partly because of the poor quality of information that reaches the public from government officials, scientists, and the news media, people may believe they are safe from environmental problems when they may in fact be at great risk.... Conversely, people may believe they are in imminent danger, when they actually are relatively safe.
>
> Should such an unfounded fear be counted as a risk to Vermonters' quality of life? The Advisory Committee said no, it would not rank a problem higher just because Vermonters are afraid of it. Nor would it rank a problem lower just because Vermonters were indifferent to it. Rather, the committee concluded that it should stress the scientific estimates of risk. That decision and this report are part of the committee's effort to help Vermonters better understand the risks around them. The Advisory Committee believes that a broader public understanding of risk will help Vermonters use their financial and political resources to reduce the most risks, rather than to allay the most fear. (VANR 1991)

Although the analysts' reports on peace-of-mind effects didn't affect the rankings, the information demonstrated to readers that the committee was aware of their fears, and it provided a useful guide for developing risk management strategies, which had to be responsive to the public's understanding of problems and willingnessto act on them. The advisory committee kept these "manageability" issues out of its rankings.

Economic Well-Being. Vermont approached this criterion much as EPA had in *Unfinished Business* and the states of Washington and Colorado had in their welfare analyses. Competent economists attempted to capture the costs that each problem was creating through increased health care and lost productivity, lost agricultural crops and recreational opportunities, degraded real estate values, and damage to building materials, fabric, and property. To put a price tag on water pollution, for example, Vermont's analysts went so far as to calculate the energy costs associated with Health Department orders to people to boil their water ($212,500 per year) and the premium people were paying to install home water-filtration systems ($2 million to $4 million per year).

These conventional techniques didn't satisfy the economists, the Quality of Life Work Group, or the advisory committee. In only a few cases were the economists willing to add up their damage estimates for a problem and present it as a bottom line. The project published no table comparing costs for all twenty problems (or inviting people to add them up for a "total cost of environmental damage"). That table was omitted for precisely the same reason that there was no table of "total expected fatalities" in the report: the analysts simply didn't believe that their numbers would convey an accurate picture of reality because so much of the picture couldn't be filled in. They could not, for example, calculate the economic impacts of wetlands loss or the invasion of the state's lakes by Eurasian milfoil.

Most of the participants were philosophically opposed to trying to put a dollar value on aesthetic damage, the extinction of a species, or other losses that don't bear a market price. Instead, the analysts tried to characterize these losses in more direct terms: the number of miles of view lost to haze over the course of a summer, for example.

Washington and Colorado also had thrown out or played down most of the economic analyses they commissioned for their projects. The participants not only lacked confidence in the numbers, but also feared that the numbers would drive all subsequent decisions and that the familiarity and apparent simplicity of dollars would make it too simple to compare dissimilar problems and dissimilar risks. The early rhetoric about CRA was that "risk" could be a "common metric" for comparing environmental problems. Practitioners discovered that no single metric—not even dollars—would suffice. Seen in this light, Vermont's quality of life analysis was an attempt to organize relevant data on as few different scales as possible, but no fewer (Minard 1993).

No Analytical Criteria: Michigan's Approach

Unlike any of the other projects to date, the Michigan team chose to not adopt specific analytical criteria and to not organize data on any predetermined set of scales. One of the project's unpublished working papers instructed the authors of the white papers to discuss the risks the problems posed: "These risks could include some combination of the following impacts: ecological; economic; human health; natural resource; social." The authors in return highlighted what they thought was most important or what they knew the most about. Thus some reports included economic damage estimates or predictions for the cost of risk management options, others didn't. William Cooper, the Michigan aqautic ecologist mentioned earlier, believed that by leaving the criteria undefined, the ranking would be more robust and less likely to convey a false sense of technical precision. The three committees went into the ranking session satisfied with the information they had in hand. EPA representatives, however, were distressed by the lack of analytical rigor (Minard, Jones, and Paterson 1993).

The virtues of Michigan's approach, like the appropriateness of contingent valuation methods, is always a good conversation starter at a gathering of comparative risk analysts. So, too, are the validity of rat studies, definitions of ecoregions, dioxin, and electromagnetic fields. The controversy that surrounds each one of these issues is just a reminder that analysts don't have the answers and that the whole enterprise of environmental management, let alone comparative risk, is fraught with uncertainty. Most practitioners agree that more people should get involved in this debate and that framing it is one of the principal values of their work.

RANKING RISKS, RANKING PRIORITIES

The state projects have treated "the ranking" in three fundamentally different ways. Colorado and Vermont ranked the problems in order of seriousness of the the risks they posed. Washington and Louisiana ranked the problems in the order in which state government should address them. Michigan did some of both. In all five cases, the states created an "integrated" ranking that aggregated all the data along all the analytical criteria. Few of the EPA projects have taken that value-laden step. These distinctions map out the zone where risk assessment and risk management converge.

In most of the state projects, risk analysts worked in teams to analyze a set of problems along a single dimension. And, in most projects, the analysts ranked the problems along that dimension. The conclusion of *Unfinished Business*, for example, was a set of four rankings: cancer, noncancer health effects, ecosystems, and welfare. In most of the state projects, each work group made a similar ranking along its own single dimension. Some of the projects used scoring systems to make it easier to add up and compare problems. Others used a discussion-based approach to sort the problems into higher and lower categories. Rather than assign scores to components of the problems, the committees debated the meaning of their findings in order to reach a consensus on a ranking of the problems. EPA texts, including early guidance to state practitioners, called these products of technical committees "technical rankings," suggesting that they were relatively objective and value-free, if not judgment-free. As the health subcommittee of the SAB demonstrated, however, even a "single" dimension of the analysis, such as human health effects, is composed of so many strands that *any* ranking is, in fact, enormously value-laden.

Scientists are capable of making thoughtful and useful rankings, but no one should infer that their rankings are inherently more "scientific" or technical than those of any other well-informed group.

Two Types of Ranking

Generally, state projects have charged their steering committees or public advisory committees with the more obviously value-laden work of making an integrated ranking: combining health, ecosystems, and a wide array of welfare effects. Most of these committees talked their way through the ranking: debating the reasons for calling a given problem a higher or lower risk. Some of the projects (such as Louisiana) insisted on consensus in this process; others (Vermont) made decisions by majority vote and highlighted individuals' reasons for disagreeing. In Louisiana, the members voted by secret ballot in a way that let staff tabulate the scores to produce a ranking. The committee then convened to agree as a group where to draw the lines separating high from medium and medium from low.

In Louisiana and Washington, the ranking included another dimension: manageability. Washington Environment 2010's Public Advisory Committee used the following criteria for its consolidated ranking, according to an undated project document:
- relative human health risks
- risks to ecological systems
- potential for causing economic damages
- apparent trend of the threat and associated risks

- manageability of the threat in terms of:
  —existing public awareness
  —existing legal authority
  —existing control programs
  —availability of control technology or techniques
  —effectiveness of control technology or techniques
  —cost of control technology or techniques
- personal and professional experience and judgment

Thus the ranking was designed to set priorities for action: state government should go after the highest risks it could manage.

This approach is intellectually and politically valid, but two problems tend to flow from it, and did in Washington. When the project published its *State of the Environment Report* in 1989 (WDE 1989), the ranking chart (in Table 1) ambiguously referred to these as "the Top 23 Threats to the Environment," implying that they were the worst problems, not simply the worst that the state could do something about. That confusion later worked its way into another 2010 document, *A Citizen's Guide to Washington's Environment* (WDE 1990a), which used the same list, implying that cost, manageability, legal authority, and so forth would apply to individuals in the same way it did to the Department of Ecology.

The Vermont committee's discussion of affronts to "peace of mind" stated the rationale for keeping manageability factors out of the ranking. The committee wanted to identify the most serious risks and leave it to individuals and officials at several levels of government to decide how to respond. That choice had a profound impact on the final ranking. Few members of the committee believed that Vermont's state government should put much effort into tackling global climate change or ventilation problems inside people's homes, but they wanted the public to know that these were high-risk problems that someone—probably the federal government and individuals, respectively—would need to deal with.

Seattle's project handled this point smoothly by doing both: ranking risks and then ranking priorities for city agencies. The project published its two rankings side-by-side to help illuminate the choices involved (City of Seattle 1991).

Interpretations and Misinterpretations of Rankings

Casual observers of the state and federal comparative risk projects have often failed to distinguish between rankings of risk and rankings of priorities. The mistake would be very serious if anyone took the rankings—the simple lists—very seriously.

In effect, the rankings—the lists—are simply the headlines on the top of a much richer, more detailed, and more useful story. They convey some information but serve primarily to get people's attention, to raise the startled observer's eyebrows. How can it be possible that indoor air pollution is a greater health threat than Superfund sites? What does this mean? How do they know? If it's true, what should I do? These questions quickly take people back to the data, to the analysis, and to a level of detail that will be necessary if they—or their state agencies—are interested in addressing the problem.

The challenge of ranking problems helps focus attention on the most important details, regardless of how the problems were organized in the original problem list and regardless of how the problem titles get sorted out in the process. For example, the Vermont advisory committee ranked "exposure to toxics in the household" in its second-highest ranking group (VANR 1991). The most significant component of the problem area was the effect of lead paint on young children, something about which the state was doing very little. These details became salient during the ranking and helped to motivate the Health Department to begin working on the problem. Those efforts helped motivate the legislature to pass protective legislation in 1993.

One of the objections to comparative risk analysis assumes that public managers and appropriations committee members are either dolts or sociopaths. The objection goes like this: once a state has ranked its risks or its priorities, managers will unthinkingly take money from the problems at the bottom and move it to problems at the top. So much money will be robbed from sewage treatment plant operations to combat radon, to draw an extreme example, that cholera will break out, making the state worse off than before, while government experts blindly insist that they must stay on the prioritized course. Surely, most public managers, legislators, and the people who hold them accountable are not dolts. Once they have started to think comprehensively about risks and risk-reduction opportunities, they should be *less* likely to make bone-headed, tunnel-visioned decisions than before. To my knowledge, there is no evidence to suggest that the projects have undermined states' capacity or willingness to keep low-risk problems low risk, even when agencies have had to cut their budgets. Washington's Department of Ecology, for instance, used its knowledge to target cuts to *minimize* their impacts, an approach that should be far superior to the more typical across-the-board squeeze.

One of the findings of NCCR's analysis of the first six state comparative risk projects was that among their most important products was a more sophisticated and cohesive staff (Minard, Jones, and Paterson 1993). The participants understood their own programs better, and they learned a great deal about their colleagues' programs and how they all interact: how, for example, a waste division's policy on the incineration of used motor oil might affect air and water quality. The state projects suggest that this educational process makes for better public management, though no one has attempted to quantify the results.

The basic belief that regulatory agencies will work more effectively if they have a core group of managers who understand the whole regulatory system and the ways different agencies manage environmental risks is at the heart of Justice Stephen Breyer's recommendations in *Breaking the Vicious Circle: Toward Effective Risk Regulation*. He concludes:

> My analysis... suggests the desirability of recruiting "coordinators" who are familiar with more than one discipline (not simply science, or law, or economics, or politics, or administration), more than one agency program, and more than one branch of government.... Such cross-branch governmental experience does not teach any basic truths about government (other than cliches); but, in the context of a particular problem (say, risk regulation) it helps create solutions that reflect an understanding of and sensitivity to the likely reactions, the points of view, and the difficulties of those institutions, including those of that ultimate source of democratic legitimacy (which Congress very well reflects), the people.
>
> Finally this book also reflects a belief that trust in institutions arises not simply as a result of openness in government, responses to local interest groups, or priorities emphasized in the press—though these attitudes and actions play an important role—but also from those institutions' doing a difficult job well. A Socratic notion of virtue—the teachers teaching well, the students learning well, the judges judging well, and the health regulators more effectively bringing about better health—must be central in any effort to create the politics of trust. (Breyer 1993)

Breyer brings the argument back to the beginning: what should public managers do when they know that they have given the public what the public has asked for—aggressive Superfund enforcement—but not what the public wants: a safe environment, peace of mind, and justice, all at a reasonable cost. Breyer is skeptical about the efficacy of public education in this context, and the state comparative risk projects do not prove him wrong: all of the projects combined have touched only a few thousand people directly, and most of those contacts have been brief.

Although the projects have not transformed the general public's perceptions of risk, they have in a few places changed enough minds to get some legislation passed. They have encouraged influential people—from line employees to governors and the administrators of EPA—to manage problems differently and, sometimes, to manage different problems. The projects have not freed any of those managers from any of the statutory or political constraints of American government.

GRANT FLEXIBILITY:
BALANCING EFFICIENCY AND ACCOUNTABILITY

EPA and two states, Colorado and Vermont, have been testing a new approach to managing risk reduction efforts that was designed to help the states address their top priorities. At the end of 1992, the states and their respective EPA regional offices agreed to conduct one-year experiments through a "cross-media demonstration grant." EPA allowed the state environmental agencies to take a small percentage of their EPA grant money and redirect it from categorical programs to problems the states had identified as high risk and high priority (NCCR 1992b).

In Colorado, the Department of Health worked with the Region 8 office to reprogram $300,000—five percent of the state's EPA grant—to invest in: multimedia coordination, compliance and enforcement; comparative risk and strategic planning; pollution prevention; integrated environmental data management; and special projects dealing with environmental lead, indoor air pollution, and groundwater wellhead protection. The state decided at the end of the first year of the experiment not to attempt to extend the pilot project because the sums involved and the degree of flexibility available were too small to compensate for the management effort.

The Vermont Agency of Natural Resources was able to create a "cross-media fund" totalling $110,000 through a combination of a special EPA incentive grant, a state match, and the redirection of unspent carryover money from two EPA program grants. The agency has been investing the fund slowly in projects that use nonregulatory approaches to address some of the highest-ranked problems in the state. Small grants, for example, have helped create a wetlands curriculum for public schools and will pay part of a new state employee's salary in 1994 to work with boaters to prevent the spread of zebra mussels in Vermont.

The modesty of these efforts underscores the real size of the problem that has governors and mayors clamoring for more flexibility to target federal money (NGA 1993). Although the stridency of some of the rhetoric coming from cities often gives their message an antienvironment, antiregulatory flavor, they have a point. The director of a drinking water system in a Midwestern city told me that EPA was going to make him spend almost $1 million a year to remove a tiny bit of lead from the drinking water. He wanted to protect kids from lead poisoning, too, but knew he could buy more protection by spending the same money on lead paint removal in public housing. "But I would go to jail," he said. After a speech for a public forum in November 1993 at the Harvard School of Public Health, in which she stressed the importance of the Clinton adminstration's health care reform efforts, EPA Administrator Carol Browner was challenged by Professor Marc Roberts: If the goal is to improve the public's health, why require municipalities to spend tens of millions of dollars to lower the concentration of radon in drinking water when the same money could buy more if invested in health care for the poor? Browner answered that the regulations were well justified by the statutes she had to administer and did not attempt to discuss the deeper issue.

Stepping back from individual program goals and requirements always seems to suggest ways to make the total package of programs work more effectively and sensibly. Environmental groups and others are justified in worrying about the misapplication of this perspective, however. Letting cities off the hook regarding the radon expenditures, for example, may do nothing to ensure that they invest the savings in neighborhood clinics, wetlands protection, or lead paint abatement.

THE NEW PROJECTS

Numerous experiments in comparative risk are under way. A few deserve a note here.

Ohio's Environmental Protection Agency is beginning work on a statewide comparative risk project, which it will conduct in loose collaboration with projects getting started in Columbus, the Northern Ohio Region (from Cleveland to Akron), and possibly Cincinnati. The efforts may help pave the way for sorting out environmental responsibilities among several layers of government.

Atlanta is attempting to use its project to better understand Atlanta's problems with environmental injustice.

In Kentucky, the executive branch is working on a comparative risk project in tight collaboration with a legislative branch research group. The project is also unusual in its handling of the time horizon. Analysts are developing several plausible scenarios of the state's future—its industry mix, economy, and population centers—and will try to relate those visions back to risk information as a tool for long-term planning.

In Seattle, under the leadership of Mayor Norman Rice, the city's planning office completed a comparative risk analysis in 1991.The staff worked with political leaders and community representatives to develop an action plan, which Mayor Rice translated into parts of his budget requests to the city council in 1992 and 1993. The results were incorporated in the city's 1994 comprehensive plan. The process has indirectly inspired the work and networks of "Sustainable Seattle," an all-volunteer effort to improve the city's future. For a time in 1994, it appeared that the city would be the first public institution to complete the cycle of comparative risk by updating the analysis and repeating the problem rankings. The departure of key staff people, however, may have stopped that initiative.

CONCLUSIONS

The strength of the comparative risk process appears to be its capacity to frame important public policy questions and to engage people in a productive attempt to answer them. Its weakness is that so many of the answers are uncertain or unwelcome or both.

As used by the states, comparative risk has added depth to policy debates and helped decisionmakers set priorities, both in times of expansion and contraction. Comparative risk results have not been rigidly applied, however. Neither ranking scores nor calculated probabilities have threatened to replace managers' judgments, politicians' instincts, or the public's right to be heard.

With or without comparative risk projects, states and cities have continued to make environmental investments in response to federal requirements and public expectations. What comparative risk projects have added to the ordinary workings of government agencies, however, has been significant. These projects have brought together scientists and laypeople, industrialists and environmental activists, bureaucrats and the people they are paid to serve, state regulators and their federal counterparts. The projects have, often for the first time, exposed these groups to each other and to each other's sources of information about the environment. The projects have often left the participants with deep new insights into their natural and political environments. Those insights and the personal associations formed during the projects continue to influence environmental management decisions from town halls to Congress. Indeed, the experiences gained in the early state and municipal projects now inform the second- and third-generation projects as the states continue to assert their own competence to set priorities and manage environmental problems.

REFERENCES

Alm, Al. 1991. Why We Didn't Use "Risk" Before. *EPA Journal.* 17 (2l): 13–16.

Bress, William. 1992. Unpublished remarks to the National Comparative Risk Practitioners' Conference, Burlington, Vermont.

Breyer, Stephen. 1993. *Breaking the Vicious Circle: Toward Effective Risk Regulation.* Cambridge, Massachusetts: Harvard University Press.

City of Seattle. Seattle Environmental Priorities Project. Technical Advisory Committee. 1991. *Environmental Risks in Seattle: A Comparative Assessment.* Seattle, Washington.

Commoner, Barry. 1994. Pollution Prevention: Putting Comparative Risk Assessment in Its Place. In *Worst Things First? The Debate over Risk-Based National Environmental Priorities,* edited by Adam M. Finkel and Dominic Golding. Washington, D.C.: Resources for the Future.

Covello, Vincent T. 1991. Risk Comparisons and Risk Communication: Issues and Problems in Comparing Health and Environmental Risks. In *Communicating Risks to the Public,* edited by Roger Kasperson and Pieter Jan Stallen. Netherlands: Kluwer Academic Publishers.

Finkel, Adam. 1990. *Confronting Uncertainty in Risk Management: A Guide for Decision-Makers.* Washington, D.C.: Resources for the Future.

Hornstein, Donald. 1992. Reclaiming Environmental Law: A Normative Critique of Comparative Risk Analysis. *Columbia Law Review* 92: 562–633.

Lash, Jonathan. 1994. Integrating Science, Values and Democracy through Comparative Risk Assessment. In *Worst Things First? The Debate over Risk-Based National Environmental Priorities,* edited by Adam M. Finkel and Dominic Golding. Washington, D.C.: Resources for the Future.

McCloskey, Mike. 1994. Problems with Inappropriately Broad Use of Risk Terminology. *The Comparative Risk Bulletin* 4(1): 3.

Minard, Richard A. 1991. A Focus on Risk: States Reconsider Their Environmental Priorities. *Maine Policy Review* 1(1): 13–28.

———. 1993. *Critical Values at Risk: Lessons from Vermont's Quality of Life Analysis.* South Royalton, Vermont: Northeast Center for Comparative Risk, Vermont Law School.

———. 1995. Comparative Risk: Adding Value to Science. In *Toxicology Risk Assessment,* edited by Anna Fan and Louis Chang. New York: Marcel Dekker.

Minard, Richard A., Kenneth Jones, and Christopher Paterson. 1993. *State Comparative Risk Projects: A Force for Change.* South Royalton, Vermont: Northeast Center for Comparative Risk, Vermont Law School.

NAPA (National Academy of Public Administration). 1995. *Setting Priorities, Getting Results: A New Direction for EPA.* Washington D.C.: NAPA.

NCCR (Northeast Center for Comparative Risk). 1992a. A Governor's Legacy. *The Comparative Risk Bulletin* 2(10): 2.

———. 1992b. EPA Flexes with Two States. *The Comparative Risk Bulletin* 2(12): 4–5.

———. 1993a. Edwards Revives LEAP Committee. *The Comparative Risk Bulletin* 3(1): 1–2.

———. 1993b. The Governors Take a Stand. *The Comparative Risk Bulletin* 3(5): 3–5.

———. 1995. Louisiana Environmental Action Plan LEAP to 2000 Makes Progress. *The Comparative Risk Bulletin* 5(5/6): 6.

NGA (National Governors' Association). 1993. The Cumulative Impact of Environmental Regulations. Policy statement adopted by the governors in February 1993.

Reilly, William K. 1991. Testimony presented before the U.S. Senate's Committee on Environment and Public Works, 25 January.

Ruckelshaus, William D. 1983. Science, Risk, and Public Policy. *Science* 221: 1026–1028.

———. 1985. Risk, Science, and Democracy. *Issues in Science and Technology* (Spring): 19–38.

Shauver, John. Michigan Department of Natural Resources. 1994. Personal communication with the author, 19 January.

Stevens, W.K. 1991. What Really Threatens the Environment? *The New York Times*, 29 January.

Thomas, Lee. 1994. Personal interview with the author, 27 January, Washington D.C.

U.S. EPA (Environmental Protection Agency). Office of Policy Analysis. 1987. *Unfinished Business: A Comparative Assessment of Environmental Problems.* Washington, D.C.: U.S. EPA.

———. Region 6. Office of Planning and Analysis. 1990a. *Comparative Risk Project Overview Report.* Dallas, Texas: U.S. EPA.

———. Science Advisory Board. 1990b. *Reducing Risk: Setting Priorities and Strategies for Environmental Protection.* Washington, D.C.: U.S. EPA.

VANR (Vermont Agency of Natural Resources). 1991. *Environment 1991: Risks to Vermont and Vermonters.* Waterbury, Vermont: VANR.

WDE (Washington Department of Ecology). Washington Environment 2010. 1989. *State of the Environment Report.* Olympia, Washington: WDE.

———. 1990a. *A Citizen's Guide to Washington's Environment.* Olympia, Washington: WDE.

———. 1990b. Compilation of Background Analyses for Action Strategies. Olympia, Washington: WDE.

4

CRA and Its Stakeholders: Advice to the Executive Office

Frederick R. Anderson

Comparative risk analysis (CRA) may well benefit from continuing methodological improvement, but the thesis of this chapter is that comparative risk methodology is already strong enough to support its widespread application. Instead, greater attention needs to be given to the current social and political context in which risk-based priority setting is addressed. This chapter discusses this context in detail, focusing on the recent developments that have brought risk methodology to the fore and examining the positions of the principal CRA players: Congress, state and local governments, environmentalists, and business and industry. Stemming from this discussion, the chapter also makes suggestions about how the Executive Office of the President might promote transparency, public involvement, and other processes that would enhance public receptivity to risk-based priority setting while responding to criticisms and concerns that have been advanced by various interested communities.

While the topic of this book is comparative risk analysis, for purposes of this chapter I have treated comparative risk analysis and risk assessment as inextricably intertwined. Improvements in risk assessment will enable better relative risk rankings to be made; increased reliance on comparative risk analysis will fuel demand for improved individual risk assessments; and criticisms of risk assessment and comparative risk analysis often are intended for both and must be treated accordingly.

Frederick R. Anderson, an attorney at Cadwalader, Wickersham & Taft in Washington, D.C., specializes in regulatory issues involving risk assessment and management, especially regulatory reviews of the potential impacts of new and existing products on health, safety, and the environment.

RISK-BASED PRIORITY SETTING: NEW WINE IN OLD SKINS

In 1987, the U.S. Environmental Protection Agency (EPA) appeared to break new ground in *Unfinished Business* (U.S. EPA 1987) by comparing its actual program funding priorities to the priorities that top EPA managers felt *should* be followed, based on health and environmental risk. Three years later, the EPA Science Advisory Board confirmed these views in the report *Reducing Risk* and went further to recommend that EPA "assess and compare the universe of environmental risks and then take the initiative to address the most serious risks, whether or not Agency action is required specifically by law" (U.S. EPA 1990). The General Accounting Office endorsed these findings a year later (U.S. GAO 1991). EPA, a mission-oriented agency, in effect criticized its own mission assignments, thereby risking both internal dissension and congressional ire.

Not long after, Stephen Breyer, then the chief judge of a federal circuit court and now a justice on the U.S. Supreme Court, also felt compelled to address comparative risk analysis but on a much grander scale than contemplated in either *Unfinished Business* or *Reducing Risk.* Breyer compared fifty-three health and safety risk regulations across eight agencies by ranking the cost per premature death averted by the regulations. He concluded that the enormous range in costs per death averted ($100,000 to over $125 million) "suggests that the entire nation could buy more safety by refocusing its regulatory efforts" (Breyer 1993). In theory, up to 1,250 more lives could be saved at the same cost by "refocusing" regulatory efforts from the costliest regulations to the least costly.

During the transition to the Clinton administration, both the Carnegie Commission on Science, Technology, and Government and the Progressive Policy Institute endorsed greater use of comparative risk analysis in decisionmaking. The Carnegie Commission report was particularly thoughtful in putting forward a series of recommendations about how the federal government, and particularly the Executive Office of the President, might better implement risk-based priority setting, and I will return to these recommendations later in this chapter (Carnegie Commission 1993). The series of essays for the new administration compiled by the Progressive Policy Institute is all the more important because of the institute's liberal credentials and close ties to the Democratic party. The institute's essay on environmental protection strongly endorses comparative risk analysis, concluding that "progressives and the new administration should support responsible risk assessment and the research on which it depends as a way to heighten public understanding, inform the debate, and improve the democratic process." Moreover, the coauthors criticized the hostility of "some environmental advocates and legislators" to the idea of weighing relative risks, concluding that their "absolutism is scientifically wrong and may prove to be politically short-sighted because it will ultimately undermine the credibility of the nation's environmental protection efforts" (Stavins and Grumbly 1993).

Not very much has happened under the Clinton administration to implement these recommendations. Vice President Al Gore's report on "reinventing government" does recommend that the federal government "rank the seriousness of environmental, health, and safety risks and develop anticipatory approaches to regulatory problems" (Gore 1993). Of some importance, an executive order (E.O. 12866) on regulatory planning and review requires as a "principle" of regulation that "each agency shall consider, to the extent reasonable, the degree and nature of the risks posed by various substances or activities within its jurisdiction." In an action-forcing mechanism, E.O. 12866 requires that each agency's Regulatory Plan include a statement with each proposed regulatory action explaining how the action will reduce risks "as well as how the magnitude of the risk addressed by the action relates to other risks within the jurisdiction of the agency" (Regulatory Planning and Review 1993).[1] Plans were afoot in the fall of 1995 to add significant new risk provisions to E.O. 12866 in an effort to steal some of the thunder of House-passed and Senate-proposed risk bills. These plans apparently fell to opposition within the agencies to universal governmentwide criteria and procedures for risk analysis.

All of this leaves the impression that comparative risk analysis has stormed virtually unheralded onto the scene since the late 1980s. The editors of *EPA Journal* wrote that while risk assessment has been around for years, "comparative risk analysis and its derivative relative risk have arrived on the scene only recently" (U.S. EPA 1993). Twenty-five years ago, Chauncy Starr laid out most of the elements of comparative risk analysis in the pages of *Science* (Starr 1969). In 1971, I participated in a two-day workshop, similar to the one from which this book grew, sponsored by the National Academy of Engineering (NAE 1972).

In 1975, the National Academy of Sciences' "Bluebook," a report of a committee chaired by Terry Davies, proposed what his committee then called a "hazard rating scheme" that would provide "a valuable tool for establishing priorities. It makes no sense to go to enormous effort and expense to reduce the amount of a chemical in the environment so as to save a few lives if a similar effort could be directed at a different chemical and result in saving thousands of lives" (NRC 1975). Almost twenty years later, we speak of comparative risk analysis and "worst-first" regulation, but we have not managed to say it better.

Breyer's ranking is instructive, as are the rankings in EPA's *Unfin-ished Business*. Yet in 1977, Henry Grabowski and John Vernon had already done for the Consumer Product Safety Commission (CPSC) what EPA did for its regulation, only the CPSC ranking is much more quantitative. The coauthors compared the CPSC's priority rankings for regulating twenty-one products (from power motors to ammonia) with the benefit-cost ratios for regulating these products, showing that a rough-and-ready correspondence existed for some priority tasks, but that for others the logic of priority, high and low, was highly question-able (Grabowski and Vernon 1977). (Power mowers, CPSC's number one priority, ranked thirteenth on Grabowski and Vernon's list. Accord-ing to the benefit-cost ratios, regulation of "bath tubs and showers" and "over-the-counter antihistamines" should rank first and second respec-tively, but the CPSC ranked them twelfth and fifteenth, respectively. Nevertheless, thirteen of the twenty-one products on the CPSC list were within five ranks of their places on the Grabowski and Vernon list, a pretty good score, given data and methodological problems.)

WHY NOW, HOW NOW? THE CURRENT EMPHASIS ON RISK-BASED PRIORITY SETTING

The intensity and variety of activity on Capitol Hill has been consider-able since 1994 and shows little sign of abating. Two additional and sig-nificant reports, one from the Congress about risk assessment research (U.S. OTA 1993) and another from the National Academy of Sciences setting out seventy recommendations for improving EPA's risk assess-ments (NRC 1994a), are undoubtedly having a major impact. What is behind all this interest in risk assessment and comparative risk analy-sis? If CRA is over a quarter of a century old, why is it surging higher on the public agenda at this particular time?

My explanation and analysis point to a series of developments that have combined to catapult comparative risk analysis higher on the agenda. In these developments, I see several common threads and con-cerns, out of which I propose to weave support for a perspective and a series of activities for the Executive Office. These common elements are developed in the next two sections.

The temptation is to explain the current level of attention to risk methodologies by pointing to a combination of forces at work.

- Over the past seven years, the leadership of the EPA has been drawing attention to comparative risk analysis.
- Congress has developed a belated concern about costs, and the resulting chaotic array of legislative requirements has been

imposed without regard to risk priority or relative expense of implementation.

- Environmentalists have rejected risk assessment and comparative risk analysis in favor of toxics use reduction and legislatively mandated bans on chemicals.
- Some social scientists and advocates of environmental equity have attempted to broaden the scope of risk assessment beyond the traditional boundaries of science.
- Industry has pleaded that we reaffirm allegiance to the common metric of risk reduction to head off an era of polarized environmental power politics where votes determine risk and substances may be legislated out of the economy without adequate knowledge of their potential for harm or their social utility.

All of these factors help push risk assessment and comparative risk analysis higher on the policy agenda, and I will address them in turn. First, however, I must note one additional factor that has been a necessary prelude to heightened attention to risk assessment, and this factor has been overlooked in the current discussions.

A Common Language of Risk

The current debate over risk assessment and comparative risk analysis would not be as important as it is, and probably would not even be possible, but for the emergence over the past decade of a science of risk assessment, a common framework of easily applied concepts and distinctions, and a useful vocabulary of risk to facilitate the public debate. (For more on the science of risk assessment, see Rodricks 1992). R. Granger Morgan has noted that "While public trust in risk management has declined, ironically the discipline of risk analysis has matured. It is now possible to examine potential hazards in a rigorous, quantitative fashion and thus give people and their representatives facts on which to base essential personal and political decisions" (Morgan 1993). Much as Leibniz and Newton enabled calculus to advance by supplying mathematical language for it, so today risk policy analysts have provided a language of risk that enables risk to move up the public agenda. The language of risk includes, not just the new analytic tools risk analysts have at their disposal, including computer-based techniques to clarify the underlying problem of uncertainty and variability,[2] but also

- the distinction, made more than a decade ago by the National Academy of Sciences in its "Redbook" report, between risk assessment and risk management, and subsequent efforts to define the interaction of the two (NAS 1983);
- the definition of "default options" or "science policy inferences" that provide a focused context for the debate about "conservative"

risk assessment, "sound" science, margins of safety, and the role of scientific judgment;
- the identification of the field of risk communication and the two-way street it requires between the public and the scientific communities;
- risk ranking through comparative risk analysis; and
- the valuable idea of "substitution risks," where side effects of risk reduction may themselves generate new risks.

Public discussion of social regulation will always be somewhat murky and difficult, because the issues are complex and occasionally very technical. But advances in structuring public discussion have made risk issues more accessible and have highlighted the social choices involved. In the recent round of congressional hearings on comparative risk analysis and in the floor debate on the rule governing the House bill to elevate EPA to cabinet status, these terms provided a more-or-less commonly accessible vocabulary for the discussion.

Particularly important have been advances in understanding issues of risk communication. The insights of Fischhoff, Lave, Morgan, Slovic, and others mean that we have starting points for a meaningful exchange between technical risk assessors, risk managers, and the public (Goleman 1994). We are poised to move beyond one-way transmission of risk assessment results to policymakers to a two-way risk analysis, where the public helps "scope out" risks to be addressed via improved risk communication. Public attention to the "risk communication problem" is itself pushing risk assessment further up the public agenda.

Most risk issues may have been present twenty-five years ago when the National Academy of Engineering conducted its ground-breaking workshop on benefit-risk decisionmaking. The state of analysis then, however, had not advanced to the point that a full public dialogue was possible. Today, that has changed. Risk specialists participating in the workshop that formed the basis for this book may see more problems than solutions; Terry Davies' long list of challenging issues to be addressed in comparative risk analysis suggests as much (see Chapter 2). But to journalists, congressional staff, their employers, and the public at large, risk issues are far more accessible than they were as recently as ten years ago.

Congress and Risk Legislation

The agendas of any Congress are set by issues the public raises for it. But recent congressional attention to risk assessment itself is generating quite a lot of public discussion beyond that stimulated by state and local resistance to "unfunded mandates" and the array of recent reports. For the first time, both the Senate and House in the 103rd

Congress were receptive to generic risk assessment reform and comparative risk approaches, as illustrated by these items:

- adoption by a wide margin of the Johnston Amendment to the Senate bill to elevate EPA to Cabinet status
- the dramatic vote in the House rejecting the Democratic leadership's attempt to block consideration of a comparative risk and benefit-cost amendment to the House EPA Cabinet bill and subsequent efforts to fashion a suitable House bill
- introduction of the Moynihan, Moorhead, and other risk bills
- proposals to require quantitative risk assessment in the Food, Drug and Cosmetic Act amendments
- major hearings in the House and Senate on risk assessment

This receptivity has crescendoed in the 104th Congress, where the House approved by a wide margin a revolutionary bill encompassing risk assessment and management, benefit-cost analysis, and regulatory reform. In the spring of 1995, the Senate appeared to be on the verge of approving a similar bill sponsored by Senate Majority Leader Robert Dole. Neither bill gave special prominence to comparative risk analysis. The long-term significance of the House and Senate actions, however, for risk-based priority setting of environmental, health, and safety items on Congress' future legislative agenda should not be dismissed as an aberration or a partisan consequence of the Republican victories in the 1994 elections.

The House, after all, passed far-reaching legislation. Senate negotiators, Democratic as well as Republican, tried to salvage a risk-benefit/cost regulatory reform bill in the fall of 1995, although chances for a Senate bill diminished as the 1996 presidential election campaign ramped up. Still, a bipartisan majority in both houses appeared willing to vote in the next year or two for "kinder, gentler," less revolutionary comparative risk assessment and management legislation with CRA components. Negotiations over Senator Dole's bill, S.343, stumbled over several problem areas: the "trigger" for full-scale assessment and benefit-cost analysis of federal regulations, the extent and timing of judicial review, and the extent of retroactive "lookback" review of existing major regulations. Most importantly, S.343 would impose a benefit-greater-than-cost test on regulations—a cross-cutting "super mandate" that would supplement and even override existing environmental, health, and safety legislation. Compromise in these issues appeared possible as of November 1995.

Proponents of risk legislation in 1995 may have asked for too much, too soon. They exposed both House and Senate proposals to criticism that the proposals would stall needed regulations and could even "turn back the clock" on two decades of environmental legislation. Proponents of risk legislation may be better advised to adopt an incremental approach, beginning with legislation that makes modest reforms in the specific ways risk assessments are performed and that requires study and comparison of relative risks. Imposing a requirement that all proposed regulations meet CRA and benefit-cost tests should wait until later: the definitions of "benefits" and "costs" in benefit-cost analysis have not yet been spelled out with sufficient clarity to permit a legislative consensus to emerge. Some powerful groups believe that the benefits of regulation will be undervalued and its costs overvalued. Similar fears about CRA exist, although contributors to this book have advanced the state of the art of CRA. None of these observations, however, is intended to detract from the fact that congressional interest in legislating risk is wide, deep, and bipartisan.

Senator Daniel Patrick Moynihan put the congressional concern in focus as early as 1993 by stating in his proposed Environmental Risk Reduction Act that, while $115 billion a year may not necessarily be excessive for the nation to spend on environmental protection, "the amount is too substantial for the funds to be used ineffectively or inefficiently" (U.S. Senate 1993). William Ruckelshaus, several years earlier, had been more pointed in his criticism of Congress, which it sometimes seems "would rather have EPA pretend to be doing 1,000 tasks than have it select the hundred most important tasks and do them well" (Ruckelshaus 1985). Today, Congress appears willing to reexamine the accumulation of three decades of piecemeal legislation to see if it reflects reasonable priorities for health and environmental protection.

Self-discipline has not historically been Congress' strong suit. Generic approaches to environmental legislation are rare, although there are some examples, such as the National Environmental Policy Act (NEPA), the Toxic Substances Control Act, and the National Forest Management Act. Also, Congress provided some precedent in the Congressional Budget and Impoundment Control Act of 1974, which created the Congressional Budget Office and a budgetary prioritization review process.

Congress' apparent concern for risk-based priority setting may give the environmental community pause as it continues to turn away from risk-based regulation and toward legislative bans and the toxics use reduction strategy. Such strategies directly contradict the thrust of the comparative risk and benefit-cost analysis movement in Congress today. Reaffirming and improving the use of risk assessment, particularly through the agencies that conduct risk-based regulation, may prove to be a more successful strategy if Congress continues down the path toward legislation to require risk-based priority setting.

State and Local Governments, Risk, and Unfunded Mandates

Local and state governments have aggressively taken the lead in promoting congressional attention to risk assessment. Yet it is clear from comparative risk "experiments" across the country, that some jurisdictions are not using CRAs merely to attack unfunded federal mandates.

In the minds of many governors, mayors, state legislators, and other local officials, the solution to the problem of "unfunded mandates" appears to be inextricably linked to better risk assessment and relative risk. In late January 1994, the National Governors' Association again called on Congress and the Clinton administration to provide federal funds for new environmental requirements imposed on states and cities. The governors also called for benefit-cost analysis to aid priority setting for environmental regulation. The U.S. Conference of Mayors called attention to Congress' unfunded mandates and adopted a resolution: "Risk assessment is a crucial and fundamental component of any cost-benefit analysis of mandates and should be included in all pending legislation which will mandate programs on state and local governments" (Inside EPA 1993). Ohio city governments, under the lead of Columbus, played a key role in drawing state and local attention to the importance of risk assessment with studies of high-cost federal environmental programs that address insignificant health risks (Ohio Municipal League 1991, 1992).

Some state and local governments have gone beyond simply criticizing federal programs that increase costs. Two dozen states and localities have attempted to make comparative risk analysis a tool for setting local program priorities (Minard, Jones, and Paterson 1993). EPA support has made some of these projects possible. What is most interesting about the results of these comparative risk experiments is that they have been used to bring the public into agency deliberations and decisionmaking, and that state and city officials have conducted the projects to try to create not just a relative risk ranking, but a process that itself helps build coalitions that favor shifting priorities to higher-risk endeavors. As noted in Chapter 3 by Richard Minard, sponsors of state CRA experiments also found that the ranking process tended to expose "weak arguments, poor data, and fuzzy thinking" and "to break down preconceptions about the problems and... prejudices about other participants," an observation that recalls how successful mediation and regulatory negotiation work. Also in Chapter 3, Minard observes that as the state and local comparative risk projects evolved, they helped comparative risk analysis become "more democratic, more inclusive, [and] more honest about its own limitations."

State and local comparative risk projects are valued because they have served to raise the level of interest in comparative risk analysis and have blunted the criticism that states and localities are cynically interested only in using comparative risk analysis as a tool for pushing their political agendas on unfunded mandates with Congress. But there is much also to be learned from them about how to organize successful federal comparative risk efforts.

The Environmental Community and Risk Methodology

The paradox of risk-based regulation is that its traditional defender, the environmental community, today routinely attacks risk-based regulation, while its traditional critic, the industrial community, generally defends risk-based decisionmaking. Environmental resistance to risk-based decisionmaking remains the central issue facing risk assessment and comparative risk analysis today. Environmentalists' problems begin with risk assessment itself, not just comparative risk ranking. Why?

In late 1993, Ellen Silbergeld of the Environmental Defense Fund described the concerns environmentalists have with risk assessment, including the following (Silbergeld 1993):

- Risk assessment implies that a certain amount of risk is acceptable, a conclusion some environmentalists believe unlawful under certain statutes and unethical in most circumstances.
- Allowing risks may mean some people will be more impacted than others, contrary to the American tradition of rejecting unequal treatment under the law.
- Better risk assessment may mean acquiring agent-specific data, such as pharmacokinetic information, which would add greatly to the burden of gathering data for regulatory purposes and may not be cost-effective for decisionmaking purposes.
- It is in practice impossible to separate risk assessment and risk management.
- Risk assessment can continue indefinitely, so adopting risk assessment as a policy tool creates difficulty in terminating data accumulation, even temporarily.
- Risk assessment can encourage the use of research as a delaying tactic.
- Actual use of risk assessment has not improved the efficiency, speed, and public acceptability of risk-based public policymaking.
- The public is suspicious of complex risk assessment and of the apparent ease with which risk assessors predict risks precisely.

Michael McCloskey, for many years the Sierra Club's executive director and now its national chairman, has written: "Use of... [risk]

terminology is part of a broader effort to obscure the role of judgment and values in pollution control and make it sound like it can be addressed solely in terms of a 'scientific process.' What is not admitted is how much of this 'scientific process' really is window dressing. However, the terminology both sounds convincing and acts to exclude the interested public from participating in the process. As such, it also disempowers the public. This emphasis on 'scientific process' also is designed to raise the threshold for action—with the demand for absolute proof before anything is done" (McCloskey 1994).

It would appear that in the eyes of many in the environmental community, risk assessment is an ethically suspect, resource-intensive, never-ending process that does not produce better decisions, that is promoted to delay regulatory decisions and can be manipulated by technical risk assessors to confuse and shut out the public and minimize their legitimate concerns, and that comparative risk analysis is used to trivialize some risks in order to ease them off the regulatory agenda.

This strong, and paradoxical, indictment of risk assessment by environmentalists can be better understood if we examine first the alternative that environmentalists offer and then the consequences to the balance of political power of environmentalists' embracing risk assessment at the present time.

Threshold Determination: An Environmentalist Alternative to Risk Assessment and CRA. The energy that environmentalists put into attacking risk methods bears a close relationship to their advocacy of an alternative that is incompatible with risk assessment and CRA in key respects. Many environmentalists today believe that instead of directing regulatory agencies to conduct careful quantitative risk assessments before regulating (much less ranking these assessments and regulating "worst first"), Congress should itself make a simple threshold determination that use of certain substances and processes should be severely curtailed or banned, with the burden on industry to show safety before use may be resumed.

The rationale for the new paradigm for environmental activism was presented with great clarity in the *State of the World* report, released by the Worldwatch Institute. The report acknowledges the present emphasis on cancer prevention and regulatory programs, but goes on to conclude that injury to the nervous, endocrine, immune, and other physiological systems may occur at much lower exposure levels than those that cause cancer. In the report, Misch argues for "a much more conservative approach to chemical regulation... perhaps through a combination of pushes and prompts such as tax incentives and federally mandated phaseouts," concluding that "overturning the presumption of innocence about chemicals is long overdue"(Misch 1994).

Perhaps it would be incorrect to say that risk assessment is not used in implementing this strategy. Congress has to debate the threat of harm presented by the substances. But Congress's "risk assessment" and the quantitative risk assessments performed under existing regulatory statutes by specialized agencies are worlds apart.

The striking thing about such indictments of risk assessment as those of Misch and Silbergeld is their impatience with careful data gathering and scientific analysis. To condemn risk assessment for not reaching the right result quickly enough may be consistent with a legislative ban and toxics use reduction approach to environmental policymaking, but in the end it undermines well-informed, science-based decisionmaking—the only foundation for health and environmental protection that will ensure public support for environmental policies over the long run.

Control of the Regulatory Agenda. The reason environmentalists resist risk assessment and comparative risk analysis appears to be concern not so much about risk assessment methods, delays, or public acceptance as about loss of control of the environmental agenda. The center of activity in risk-based regulation and priority setting is in the regulatory agencies with their scientists, their scientific advisory committees, and the scientific researchers who perform risk assessments in private organizations, whereas the center of activity for considering legislative bans and toxics use reduction measures is in the Congress, where environmental organizations have been influential with key members and their staffs. The legislative ban and toxics use reduction approach keeps environmental regulation under close political control, fulfilling McCloskey's admonition that pollution control is more a matter of "judgment and values" than of a "scientific process."

Some Environmentalist Perspectives on Current Risk Methodology. A plethora of statutes currently requires that some type of risk assessment be conducted as a basis for regulatory standard setting. With respect to these risk decisions, some environmental critics advance a theory of risk assessment and comparative risk analysis that obliterates the critical distinction between risk assessment and risk management and thereby undermines the scientific credibility of risk assessment (and thus of comparative risk analysis). The theory appears to be that if risk assessment is going to be used, at least it should be made as "conservative" as possible, to reflect the policy agenda of environmental organizations.

Howard Latin has taken this point of view to its logical conclusion (Latin 1988). Latin completely rejects the risk assessment/risk management distinction established by the National Academy of Science's "Redbook" (see NAS 1983). He challenges "the conventional view that scientific perspective should dominate the risk-assessment process" and contends instead that "social policy considerations must play as prominent a role in the choice of risk estimates as in the ultimate determination of which predicted risks should be deemed unacceptable." Latin questions the "social ramifications of EPA's current emphasis on 'good science,'" writing that "EPA's risk assessment focus is likely to result in reduced public protection against potential toxics hazards, increased regulatory decision-making costs, and expanded opportunities for obstructive behavior by agency bureaucrats or private parties hostile to toxics regulations." Citing EPA's 1986 Cancer Risk Assessment Guidelines, Latin castigates EPA for seeking "the most scientifically appropriate interpretation" in risk assessment and for attempting to ensure that it "be carried out independently from considerations of the consequences of regulatory action" (Latin 1988). (The guidelines cited by Latin are found in 51 *Federal Register* 33: 992–93.) The argument is also made that risks cannot be ranked for "worst first" regulation merely based upon the relative risks presented. Certainly, many factors other than quantitative health risk are relative to the ranking. But one type of useful information is health risk rankings and relative costs of implementing controls. To inject nonquantitative, nonscientific information into the comparative risk analysis at this stage would convey a false sense of "scientific" validity to the ranking. Other considerations are, of course, valid in managing risky endeavors, but this does not mean that comparative risk analysis should be made the vessel for these considerations.

The Environmental Justice Movement and Risk Methodology

The attention that the environmental justice movement has brought to risk assessment and comparative risk analysis also has helped push them further into the limelight. The movement maintains that low-income communities and racial minorities have been disproportionately exposed to environmental disruption through skewed implementation of environmental programs, an injustice that should be redressed in permitting and siting decisions. President Clinton, in 1994, signed an executive order instructing federal agencies to consider the socioeconomic and racial profiles of a community before issuing permits, siting approvals, or new rules. To facilitate implementation of the policy of equal treatment in environmental decisionmaking, the proposed executive order would require federal agencies to collect information about income, race, and other socioeconomic factors and to ensure that health risk assessment takes the community's profile into account—a sort of "social impact statement" (Schneider 1993, Cushman 1993). The 103rd Congress had several pending bills on environmental justice and equity. The EPA Office of Civil Rights, in 1994, opened an investigation, under the 1964 Civil Rights Act, of Louisiana and Mississippi's implementation of their environmental programs. The Sierra Club Legal Defense Fund has sued to block permitting decisions for a hazardous waste plant in Louisiana and a cement kiln, a hazardous waste treatment facility, and a landfill in Mississippi. Church leaders have formed an interfaith National Partnership for the Environment to Promote Environmental Justice (Neibuhr 1993).

Proponents of environmental justice believe that risk assessment should provide the foundation for more equitable environmental decisionmaking. Yet at the same time, they perceive risk assessment as overly technical and inaccessible to community groups. They fear exclusion from decisionmaking and mistrust risk science. The views of environmental justice advocates are still evolving, but in most discussions of risk assessment, either they argue for an expanded concept of risk assessment that takes community concerns directly into account (and thus would appear to inject risk management into risk assessment) or, more appropriately, they argue for risk assessment that includes attention to assessing the relative risk of various affected racial and socioeconomic groups.

Risk assessment can certainly be a useful tool for identification of health impacts on particular communities whose diets, genetic backgrounds, and habits suggest that particular risks be evaluated. The methods for doing so are well developed, but more could be done to ensure that the "scope" of a risk assessment includes such studies. Comparative risk analysis can also be a useful tool for identifying the differential impacts of public policies on various socioeconomic and racial groups.

Yet if risk assessment becomes a tool for making the case for disproportionate impact or for developing civil rights cases, risk assessment and comparative risk analysis will be just that much more politicized. There will certainly be pressure and the temptation to use conservative assumptions in a risk assessment affecting a low-income or minority group in order to right historic injustices.

The Risk Considerations of Industry and Business

The industrial and business communities support risk-based decisionmaking, and especially comparative risk analysis, because it is science-based and thus more likely to protect the public from harm while continuing to permit the delivery of substantial benefits, including health, safety, and environmental benefits. To the business community, risk-based decisionmaking seems much more likely than any other basic environmental strategy to provide the long-term certainty for business investment that the producers of goods and services need if the economy is to grow in stable fashion. For its part, comparative risk analysis permits prioritization among environmental programs, which promotes efficiency in government and—if not a reduced burden on industry—at least more health, safety, and environmental protection for the same level of investment in environmental programs.

In some respects, industry's support of existing risk-benefit decisionmaking and proposed comparative risk ranking appears more enthusiastic than it might otherwise be, because of pressure by environmental groups for bans and legislative burden-shifting. In the view of industry, the strategies that environmentalists have pressed on Congress today have a weak basis in science and thus are subject to reversal if the political complexion of Congress changes, with a consequent waste of investment and a need to redirect production in light of changed policy. While *State of the World* views with approval "overturning the presumption of innocence" of chemicals, industry in general sees something else: a highly experimental and wasteful attempt to shift to them the burden of proof of safety for a long, open-ended list of items. (That list includes useful chemicals that are assumed to cause harm and legislated out of the stream of commerce, where they cannot even continue to be used, much less released at safe levels.)

Safe and environmentally sound "technology-forcing" may occur for some banned substances, as it has in the past. Yet some environmental advocates—both in and out of government—have proposed that scores of chemical compounds be banned or subjected to toxics use reduction, thus incurring substantial costs to society. These costs include: the substitution risks that alternative substances may create; reductions in product safety, service, and convenience; incremental energy and materials costs; and bankruptcies. In industry's view, it appears extreme to incur these costs without careful, scientifically credible risk assessments first being performed; the cursory risk assessments and CRAs undertaken directly by Congress are a wasteful, disruptive, and "risky" way to set national environmental policy. (The chemical industry in 1994 asked President Clinton to repeal or revise two pollution prevention executive orders signed earlier that year, because they do not require EPA to base prevention on risk considerations.)

Industry continues to press for actions that would improve risk assessment in hundreds of governmental decisions, while at the same time laying the foundation for better comparative risk assessment. These improvements include:

- greater use of credible science in risk assessment (for example: replacing default options as soon as possible with specific pharma-cokinetic research findings and providing central estimates of risks accompanied by a description of probabilities of alternative outcomes)
- scientific peer review of agency risk assessments
- greater public input and opportunity for review
- attention to risk communication
- federal coordination, leadership, and consistency in risk assessment

Many of the reforms that industry seeks would improve comparative risk analysis in the same ways others would have it improved, a topic to which I give attention in the next section.

Comparative Risk Analysis: Moral Choices

The most frequently advanced justification for comparative risk analysis is economic, as some members of Congress now urge. We currently spend such huge sums on reducing risks that we cannot afford to do so inefficiently. But there are strong moral grounds for comparative risk analysis. We owe it to ourselves to confront our choices honestly.

It is just this moral dimension that accounts, paradoxically, for a final source of resistance to use of comparative risk analysis. There is something in all of us, in our desire to preserve social comity, that resists comparative risk analysis. Comparative risk analysis rubs our noses in the choices we must make to solve one problem while neglecting another, or to protect one group while turning away from helping another. Comparative risk analysis asks us to make explicit our implicit choices. We do not want to hear that scarcity is not just an economic law but is part of the human condition, that it can be avoided for some but not for all. Comparative risk analysis thus belongs with other unpleasant tasks, such as deciding whether to have surgery or writing a will. When forced to gaze on our choices, we want to avert our eyes, to avoid facing the moral consequences of the implicit and explicit decisions made. For certain choices—what Guido Calabresi and Phillip Bobbitt aptly called "tragic choices" (Calabresi and Bobbitt 1978)—strong feelings may be mobilized: choices involving lifesaving technologies; life-taking activities such as military service; and perhaps even choices to regulate to protect life, health, or the environment. Yet we have a moral obligation to confront the fact that our implicit choices to protect or leave unprotected, to spend or not to spend, and to allocate scarce resources to less productive uses have consequences just as explicit choices do. Comparative risk analysis, by definition, calls on us to face up to the moral implications of these choices.

Conflict or Consensus? The Status of the Risk Debate

From the discussion of the current level of interest in risk assessment, several conclusions emerge that are directly relevant to the strategies that the Executive Office should adopt in response:

1. Congress is very interested in enacting legislation on risk assessment, comparative risk analysis, and benefit-cost analysis.
2. Agency receptivity to comparative risk analysis for budgetary priority setting remains at an all-time high, particularly at the EPA.
3. As a threshold problem, risk-based priority setting and risk assessment are under attack by environmental advocates who favor legislative strategies based on chemical bans and toxics use reduction.
4. Environmentalists also resist ceding their strong influence on the environmental agenda in Congress to priority setting by agencies and scientific organizations.
5. Nevertheless, despite the debate over basic environmental strategies, agency decisionmakers depend upon risk assessments now more than ever before, the importance of risk assessment to decisionmaking has become more widely understood by the public at large, and interested groups have increasingly sought to have their particular concerns included in the risk assessment process.
6. As one consequence, strong pressures exist to drive risk management considerations back "upstream" into risk assessment and comparative risk analysis.
7. Still, some groups fear comparative risk analysis and risk assessment because they do not understand the quantitative methods employed, do not trust the results presented, and feel excluded from the process that produced the risk assessment ranking.
8. Objections to risk assessment and comparative risk analysis based on data needs, technical content, and delays seem to follow from prejudgment that a need to regulate exists, while objections that stress tactical delay and exclusion seem well-founded but answerable via improved risk assessment processes.
9. The objection that comparative risk analysis may result in reduced environmental protection against "trivialized" risks and in a reduced environmental share of the federal budget "pie" seem misplaced, since comparative risk analysis offers a basis for stronger industry-environmental consensus on risk reduction and federal budget priorities.
10. Creative risk projects on the state level provide encouraging evidence that consensus-building risk-ranking exercises are possible and offer valuable experience upon which the federal government can build.

11. Despite their significant differences, all groups interested in risk assessment and comparative risk analysis appear to seek similar things: access to, and proactive participation in, risk assessment and comparative risk analysis processes that produce assessments and rankings they consider fair.

12. No insuperable barriers to reliance on risk assessment and comparative risk analysis appear that cannot be overcome by appropriate public processes and safeguards.

The manner in which these conclusions might be acted on is the subject of the next section.

FROM THE TOP: RISK-BASED PRIORITY SETTING IN THE EXECUTIVE OFFICE

While comparative risk analysis has been discussed for many years, it has lately risen much higher on the public agenda. The discussion of the previous sections explored how comparative risk analysis and risk assessment came to attract such a high level of recent attention. In light of that discussion, what can be done to obtain the benefits of comparative risk analysis and risk assessment while addressing their perceived deficiencies? More particularly, what could the Executive Office of the President do? A few ideas follow.

The underlying assumption of the five following recommendations is that the Executive Office can do a great deal to provide coordination, leadership, and uniformity in the use of comparative risk analysis throughout the federal government. Some critics both inside and outside of the agencies fear direct control by the Office of Management and Budget (OMB) and the Office of the President if the Executive Office plays on enhanced central leadership role. The way lies open, they suspect, for political manipulation of regulatory decisions through abuse of the risk assessment process. To some extent, this criticism must be met head-on, as Breyer (1993) suggests, because insulation of politically responsible officials from major policy decisions is not desirable and, in the end, not practically possible. Yet something can be done, not by creating an enlightened and compassionate technical elite, but through the processes I recommend below that focus on risk communication and public input, transparency, and safeguards against possible abuse.

- Insist on credible, neutral science as the necessary foundation for comparative risk analysis.
- Review the possibility that legislation may be necessary before risk-based priority setting can be effectively used in the agencies.

- Require that risk assessment and comparative risk processes are based on openness, early involvement of affected interests, and consensus building.
- Implement comparative risk analysis incrementally, beginning with agency site-specific and intraprogram comparative risk analysis.
- Promote impartial comparative risk analyses and avoid the appearance of promoting a particular regulatory agenda.

Science as CRA's Foundation

Insist on credible, neutral science as the necessary foundation for comparative risk analysis. As any comparative risk project moves forward, it will increase in visibility and importance. Perhaps understandably, various groups will exert pressure to broaden its scope and include concerns that go far beyond a ranking of health and environmental risks based on quantitative risk assessment. Concerns related to how risks might be tackled in rank order relate to risk management and do not belong in an objective, scientific ranking of risks.

Most advocates of comparative risk analysis view it as but one tool—albeit an important tool—to be employed in making policy decisions about which risks to tackle first. Risk rankings based on scientific risk assessment do not take account of any number of factors relevant to a final decision (voluntary vs. involuntary risk, public dread of dramatic or rare catastrophic events, uneven impact on different populations, and so forth). Yet pressure will arise to push these concerns upstream into the ranking process itself, that is, to turn comparative risk analysis into a comprehensive review of environmental priority setting.

Yielding to this pressure will overburden the comparative risk process and drive it afield of its scientific basis. What science can offer to the policymaking process is a quantitative expression of relative number of lives lost (or saved), injuries caused (or averted), workdays lost (or saved), and so forth, and a quantitative expression of the comparative costs of implementing controls that reduce risks. Beyond this, comparative risk analysis enters on softer, more qualitative ground.

Some comparative risk projects, particularly on the state level, attempt to poll public perceptions of risks and the rank order in which regulators should address them. Comparative risk analysis may provide a useful input to such projects, but the term is stretched to the breaking point when applied to outputs of such a project. While worthwhile as inquiries into public perception of risks, these efforts appear to go beyond the province of science. They have a role to play in setting priorities for action, but that role is not as a source of objective data on risks incurred.

This is not to say that the public has no role in the preparation of a comparative risk analysis or a risk assessment. On the contrary. Richard Minard has described in Chapter 3 the role in improving upon poor data and dispelling fuzzy thinking that comparative risk analysis may achieve. In the recommendations below I sketch a role for the public in assuring that comparative risk analysis and risk assessment are accurate, useful, and transparent. But this does not mean that comparative risk analysis should reflect the "views" on risk of affected groups, except as such. This conclusion, however, does leave open an important question about how to delineate risk assessment from other factual inquiries relevant to sound risk management.

Legislation as a Possible Prerequisite

Review the possibility that legislation may be necessary before risk-based priority setting can be effectively used in the agencies. The optimum use of comparative risk analysis is as a tool to help agencies rebudget to get more health and environmental protection for the regulatory dollar. Yet budgetary constraints requiring spending on mandated programs prevent full use of comparative risk analysis, and problems exist with using comparative risk analysis without a statutory basis for doing so.

The EPA Science Advisory Board's recommendation in *Reducing Risk*—that EPA "take the initiative to address the most serious risks, whether or not agency action is required specifically by law" (U.S. EPA 1990)—may seem like sound advice from a scientist's or program manager's point of view. The implication, however, that comparative risk analysis may require an agency to override its existing statutory mandates has been criticized by some law professors and members of Congress. The desire to act despite lack of statutory directives also applies to EPA's recent attempts to cure statutory deficits through enforcement action settlements that include pollution prevention measures and "supplemental environmental projects" for which the statutory basis may be doubtful or nonexistent. Congressman John Dingell in particular has challenged this practice.

As Professor Blomquist has argued, "it is one thing for the Science Advisory Board to recommend to the EPA that it engage Congress and the President in a spirited debate over environmental spending priorities and legal norms.... It is, however, something quite different to unilaterally shift agency priorities notwithstanding contrary legislative or executive orders detailing priorities." The Science Advisory Board advice "seems to ignore or downplay the importance of rule of law values such as certainty, predictability, reliability, and evenhandedness in our society" (Blomquist 1991).

Without a statutory mandate to use comparative risk analysis, agencies such as EPA may have very little "wiggle room" to apply comparative risk analysis effectively. At present, EPA use of comparative risk analysis affects only the margins of funding. In 1992, then-Deputy Administrator Henry Habicht said that only 5–10% of EPA's budget is a candidate for risk-based priority setting (Habicht 1994). In fiscal year 1993, Congress added about 100 specific items to EPA's responsibilities but approved almost no budget increases over fiscal year 1992 (Cleland-Hamnett 1993). A study of planning in environmental organizations for the Carnegie Commission confirmed that little budget flexibility existed to apply risk-based setting inside EPA or any of the other environmental or health agencies (Andrews 1993).

In this state of affairs, agency and Executive Office use of comparative risk analysis may be limited to application as a tool to critique existing program performance and, armed with these conclusions, "to engage the Congress and the President in a spirited debate" over priorities and proposed revisions in environmental laws. With appropriate legislation, however, enabling them to shift budget resources between programs despite statutory deadlines and mandates, subject to appropriate standards, much more could be achieved through comparative risk analysis. The Office of Science and Technology Policy (OSTP) and OMB should consider what congressional action might be necessary to permit greater use of comparative risk analysis. Perhaps this could be part of the review of risk legislation and prospects for greater use of comparative risk analysis by agencies already assigned to the Committee on Risks under the Regulatory Working Group established at OMB in late 1993.

Openness, Involvement, and Consensus Building as Requirements

Require that risk assessment and CRA processes are based on openness, early involvement of affected interests, and consensus building. At Resources for the Future's 1992 Annapolis Conference on Risk Assessment, Alice Rivlin challenged EPA to develop a priority-setting system that avoids expert elitism and encourages public participation and empowerment (Rivlin 1994). The Executive Office could do a lot to promote comparative risk analysis and to answer the criticisms discussed above, if it were to focus short-term efforts on a CRA dialogue and on establishing appropriate processes for risk assessment and comparative risk analysis, rather than on publishing actual rankings of relative risks. The pressure on the agencies and the Executive Office to produce results will be great but should be resisted in the short run. Risk communication, ventilation, access, and persuasion will help bring acceptance for the project ahead.

The Executive Office has made a good start. In the fall of 1993, the informal Executive Office risk policy group invited the agencies that regulate risks to discuss risk assessment issues with them. One of the purposes of the meetings apparently was to try to get agreement to work toward a comprehensive and consistent federal approach to risk. From all reports, calling the agencies in early did ease agency fears that OMB again might be attempting to dictate policy to them.

The Executive Office planned to continue these discussions and ultimately to produce a series of "guidance memoranda" to the agencies, beginning with a memorandum on basic approach and philosophy and moving on to subsequent memoranda as necessary on risk assessment, management, communication, and priority setting. But the intense congressional debate in 1994–95 appears to have deflected Executive Office attention from preparing these memoranda. Instead, as a counter to the congressional bills, the OSTP and OMB have floated the idea recently (late 1995) of a comprehensive executive order.

To promote greater public involvement, scientific peer review, and consensus, the Executive Office might take another look at processes that have worked well under the Administrative Procedure Act, the National Environmental Policy Act, and regulatory negotiation. Perhaps the Administrative Conference of the United States could assist the Executive Office in designing appropriate mechanisms.

The National Academy of Sciences has conducted an exercise that could provide particularly useful guidance for the Executive Office risk effort. At the request of Department of Energy Assistant Secretary Thomas Grumbly, the National Research Council convened a "Committee to Review Risk Management in the Department of Energy's Environmental Remediation Program." The committee's report, *Building Consensus* (NRC 1994b), presents a number of approaches that the committee felt would prove useful in developing support for risk assessment and comparative risk analysis in the problem-plagued program at the Department of Energy (DOE). To practice what it preached, the committee conducted a workshop of representatives of state agencies, Native Americans, affected local communities, environmental groups, DOE contractors, labor unions, and scientists. The workshop was observed by 300 individuals and was broadcast to 300 additional sites via satellite.

Table 1 contains a figure from the committee report summarizing suggested procedures for risk assessment at DOE remediation sites. The summary is instructive (NRC 1994b). It suggests:
- an initial stakeholder identification phase to identify the array of interests affected by risk assessment,
- a "scoping" phase modeled on Council on Environmental Quality (CEQ) guidelines requiring scoping for NEPA environmental impact statements,

- multiple reviews of the ongoing process by expert external scientific advisory panels,
- reviews of work plans and risk assessments reflecting the draft-final comment process of NEPA,
- multiple opportunities for consensus-building, and
- other measures intended to restore confidence in risk assessment and risk-based priority setting in the DOE remediation program.

Note that while the NRC committee's recommendations focus on public acceptance, unlike some of the state "experiments" with comparative risk analysis, the committee's suggested procedures consistently maintain the distinction between risk assessment and risk management. Thus, for example, the scoping exercise provides for identification of "community-specific scenarios (e.g., dietary practices)," "non-traditional roots of exposure (due to local customs, culture, and population)," and "preliminary data-gathering," such as anecdotal data from the "public on health and environmental impacts of concern due to exposure." The risk assessment, however, remains a scientific inquiry throughout and does not become a procedure for selecting risk management options.

Incremental Implementation of CRAs

Implement comparative risk analysis incrementally, beginning with agency site-specific and intraprogram comparative risk analysis. One of the chief criticisms leveled at CRAs is that it is impossible to develop a common metric for comparing the tremendous variety of health and environmental risks that current legislation addresses. A matrix of common measures of comparability may eventually evolve that permits more exact, if not ideal, comparisons among various types of health and environmental risks. In the meantime, comparisons of such things as lives saved, injuries averted, ecological damage avoided, and dollars saved or lost seem already to be quite useful. The comparative risk project could run aground, however, if it attempts too much at the very beginning.

Comparative risk efforts are more useful the more they are targeted on commensurable outputs. Thus agency intraprogram comparisons are most useful, intra-agency comparisons are also useful, interagency comparisons are useful but present certain challenges, and comparisons across the entire federal budget, if the late Aaron Wildavsky was correct about the federal budget process, are of limited practical utility.

It is helpful to imagine an inverted pyramid of risk-benefit priority setting. At the inverted apex are site-specific and intraprogram rankings. As the pyramid expands, and comparisons occur across the full spectrum of possibilities—including comparisons with priorities that

Table 1. Suggested Procedures for Performing Risk Assessments at DOE Remediation Sites

Step	Action	Activities
1	Stakeholder identification	Identify stakeholders in the risk assessment process (e.g., local residents; federal, state, and local citizen groups; federal, state, and local environmental groups; native american governments and associations; workers, unions, industry, and other economic interests; federal, state, and local environmental, safety, and nuclear regulatory agencies; local, county, and state government; universities and research groups; technical advisors and reviewers)
2	Scoping	Identification of: potential human and nonhuman targets potential adverse consequences (cancer, non-cancer, and ecological) general risk assessment methods perspectives and values of stakeholders (historical, aesthetic, religious, and cultural) high-risk groups community specific scenarios (such as dietary practices) nontraditional routes of exposure (due to local customs, culture, and population) monitoring needs relevant to specific exposure concerns information needs and data quality objectives Coordinate: regulatory requirements and timetables Consider: possible future land uses; possible exposure during that use Conduct: preliminary data gathering (such as historical records on occurrence and patterns of disease and population activities; anecdotal data from public on health and environmental impacts of concern due to exposure; and particular ecological impacts of concern)
3	Scoping Review	Review scoping by external national advisory panel of scientific experts
4	Work Plans	Establish items to be addressed in risk assessment Identify contaminants, pathways of exposure, and end points of concern to stakeholders Specify methods to be used in assessing exposures and risks Identify research needs relative to data gaps and methodological needs Identify kinds of monitoring needed to improve risk assessment Outline timetable for assessment and remediation activities

Table 1. *Continued*

Step	Action	Activities
		Determine mechanisms and timing of public participation
		Indicate how results of risk assessment and remediation activities will be provided to public
		Determine mechanisms to provide technical and scientific assistance to public
		Develop list of options for site remediation and future land use
		Indicate how stakeholder perspective and views are to be addressed
5	Review of Work Plans	Review scoping by external national advisory panels
6	Preliminary Risk Assessment	Conduct site-oriented preliminary risk assessment
		Identify data gaps
7	Risk Assessment Review	Review risk assessment by external national advisory panels
8	Step I Risk Assessment	Conduct assessment (as needed) for long-term risks
		Conduct further site characterization (as needed)
		Conduct off-site exposure monitoring and assessment
9	Risk Assessment Review	Review of risk assessment by external national advisory panels
10	Site Remediation and Future Land Use Option Selection	Determine (with public participation) options for site remediation and future land use based on results of Step I risk assessment
11	Remediation	Conduct remediation activities
		Monitor exposure to environment, population and workers
12	Remediation Monitoring Review	Review monitoring plan and results by external national advisory panels
13	Step II Risk Assessment	Determine effect of remediation activities on workers and public as data are available
14	Risk Assessment Review	Review remediation risk assessments by external national advisory panels
15	Population Monitoring	Continue long-term population exposure measurements and surveillance following remediation (where justified)
		Review monitoring plans and results by national advisory panel of scientific experts
16	Step III Risk Assessment	Update risk assessments on periodic basis until site no longer poses potential hazard to population or environment
		Review risk assessment implementation and results by national advisory panels

Source: NRC 1994b.

lie outside the environmental area (transportation, defense, housing, and so forth)—comparative risk analysis encounters increasing difficulty of commensurability. The Executive Office should focus initially on setting up comparative risk efforts where commensurability is the least difficult, to enhance the credibility of comparative risk analysis. The temptation should be resisted, at least initially, to apply comparative risk analysis broadly in congressional discussions of priority setting across the entire ambit of federal legislation.

To integrate comparative risk projects "vertically" on the inverted pyramid of comparative risk analysis, it may be useful to draw upon the concept of "tiering" environmental impact statements as found in CEQ guidelines. The CEQ has worked through a more-or-less consistent approach to integration of a variety of impact statements (site-specific, segment, geographic/regional, and programmatic). The tiering concept, and the public scoping exercise that integrates the different levels of impact statements, provide a useful model for comparative risk analysis.

Impartiality and Specificity of CRAs

Promote impartial CRAs and avoid the appearance of promoting a particular regulatory agenda. Comparative risk analysis had early origins in the attempts of nuclear power advocates to show that the risk of nuclear plants was small compared to risks routinely faced by the public at large. (See NAE 1972.) This early advocacy context continues to taint comparative risk analysis today. Many of its advocates seem to be saying that identification of regulation that would permit large risk reductions at low cost *compels* the conclusion that resources should be transferred to them from low-risk regulations, or that the same comparison *compels* the tolerance of the former as "acceptable risk." Yet several thoughtful CRA proponents have discovered that public interest in and acceptance of relative risk comparisons are much higher when done only to present a particular risk in context and gain perspective on the size of the risk (See Slovic, Kraus, and Covello 1990, as well as the articles cited therein).

This observation may seem pointless: What good is comparative risk analysis if we are not going to use it? But a more complex, multistep public policy formulation process may be necessary before comparative risk analysis can be fully utilized. Risk communication, discussion, and consensus-seeking—all short of budgetary decisionmaking—may be necessary before comparative risk analysis actually brings about real change in public policy. A comparative risk project in the Executive Office that clearly has no predetermined agenda may thus have a greater long-term impact on risk-based decisionmaking. Besides, most advocates of comparative risk analysis concede in the end that it is but *one* tool to be used in national priority setting, alongside statutory imperatives and other policy considerations.

CONCLUSIONS

The chapters in this book present constructive approaches to greater use of comparative risk analysis, but they also identify a large number of difficulties with carrying out comparative risk analysis. Terry Davies' list of issues to be addressed (see Chapter 2) is in some respects daunting. Some critics appear to overstate the criteria that successful comparative risk analysis should meet and demand that they be met, making the best the enemy of the good.

I am not sure whether methodological improvements are a necessary prelude to greater use of comparative risk analysis. I am not an expert in the technical methodologies of risk assessment, and my suggestions have focused on process changes that could bring greater acceptability to comparative risk exercises, on the theory that much of the reluctance to employ comparative risk analysis is based upon how the exercise may affect priority setting to the detriment of some groups and may shift the balance of power among Congress, the agencies, and science.

As a student of Washington policymaking and agency behavior for more than twenty-five years, I do have a final observation about the efficacy of attempts to employ finely tuned analytic techniques in service of public policy. They rarely work well. But the basic concept of risk-based priority setting, even if not done in an ideal fashion, is sound. Well-informed analysts of current environmental policies are convinced that we are not making optimum use of our regulatory resources to obtain a greater measure of health and environmental protection. The objection that we cannot use comparative risk assessment with great precision to reallocate our efforts should not defeat its use, even if as a somewhat crude tool. However imprecise, there are few if any other tools now available to do the necessary job of setting priorities for refocusing our environmental agenda.

ENDNOTES

[1] E.O. 12866 also endorses a "kinder and gentler" requirement for benefit-cost analysis than that pursued in the Reagan and Bush administrations. See

both E.O. 12291 in 3 C.F.R. 127 (1982) and E.O. 12498 in 50 Federal Register 1036 (1985). The "principles" section of E.O. 12866 emphasizes how agencies should assess "all costs and benefits and select those approaches that maximize net benefits, incorporating economic and distributive impacts and equity considerations into the analysis. Agencies should adhere to the principle that the regulations should be pursued in the most cost effective manner, the benefits of the regulation should justify these costs, and the regulation should be based on the best scientific, technical, economic, and other information." E.O. 12866 affirmatively recommends the adoption of economic incentives as a substitute for direct regulation, as well as requiring comparative risk analysis as mentioned above.

[2] "Only a few years ago... detailed study of risks required months of custom programming and days or weeks of mainframe computer time. Today a variety of powerful, general-purpose tools are available to make calculations involving uncertainty.... These programs, many of which run on personal computers, are revolutionizing the field.... Although using such software requires training, they [sic] could revolutionize risk assessment and make rigorous determinations far more widely available" (Morgan 1993).

REFERENCES

Andrews, N.L. 1993. Long-Range Planning in Environmental and Health Agencies. *Ecology Law Quarterly* 20: 515.

Blomquist, Robert F. 1991. The EPA's Science Advisory Board's Report on "Reducing Risk: Some Overarching Observations Regarding the Public Interest." *Environmental Law* 22: 149–88.

Breyer, Stephen. 1993. *Breaking the Vicious Circle: Toward Effective Risk Regulation.* Cambridge, Massachusetts: Harvard University Press.

Calabresi, Guido, and Phillip Bobbitt. 1978. *Tragic Choices: Conflicts Society Confronts in the Allocation of Tragically Scarce Resources.* New York: W.W. Norton.

Carnegie Commission on Science, Technology, and Government. 1993. *Risk and the Environment: Improving Regulatory Decision Making.* New York: Carnegie Corporation.

Cleland-Hamnett, Wendy. 1993. The Role of Comparative Risk Analysis. *EPA Journal* 19(1): 18–23.

Cushman, John H., Jr. 1993. U.S. To Weigh Blacks' Complaints About Pollution. *New York Times*, November 19, A16.

Goleman, Daniel. 1994. Assessing Risk: Why Fear May Outweigh Harm. *New York Times*, 1 February, C1.

Gore, Albert. 1993. *Creating a Government That Works Better and Costs Less: Report of the National Performance Review.* 168 (Recommendation REG 07). Washington, D.C.: U.S. Government Printing Office.

Grabowski, Henry, and John Vernon. 1977. Consumer Product Safety Regulation. Proceedings of the American Economic Association's annual meeting. *American Economic Review* 68(2): 284–89.

Habicht, F. Henry, II. 1994. Rationalism and Redemocratization: Time for a Truce. In *Worst Things First? The Debate over Risk-Based National Environmental Priorities*, edited by Adam M. Finkel and Dominic Golding. Washington, D.C.: Resources for the Future.

Inside EPA. 1993. Untitled news item about the U.S. Conference of Mayors. 14(26): 5

Latin, Howard. 1988. Good Science, Bad Regulation, and Toxic Risk Assessment. *Yale Journal on Regulation* 5(1): 89–148.

McCloskey, Michael. 1994. Problems with Inappropriately Broad Use of Risk Terminology. *The Comparative Risk Bulletin* 4(1): 3.

Minard, Richard, Kenneth Jones, and Christopher Paterson. 1993. *State Comparative Risk Projects: A Force for Change*. South Royalton, Vermont: Northeast Center for Comparative Risk Assessment, Vermont Law School.

Misch, Ann. 1994. Assessing Environmental Health Risks. In *State of the World*, edited by Lester Brown. (Tenth annual report.) Washington, D.C.: Worldwatch Institute.

Morgan, R. Granger. 1993. Risk Analysis and Management. *Scientific American* 269(1): 2–41.

NAE (National Academy of Engineering). Committee on Public Engineering Policy. 1972. *Perspectives on Benefit-Risk Decision Making*. Washington, D.C.: National Academy Press.

NAS (National Academy of Sciences). Committee on the Institutional Means for Assessment of Risks to Public Health. 1983. *Risk Assessment in the Federal Government: Managing the Process*. Washington, D.C.: National Academy Press.

NRC (National Research Council). Committee on Principles of Decision Making for Regulating Chemicals in the Environment. 1975. *Decision Making for Regulating Chemicals in the Environment*. Washington, D.C.: National Academy Press.

———. Committee on Risk Assessment of Hazardous Air Pollutants (CAPRA). 1994a. *Science and Judgment in Risk Assessment*. Washington, D.C.: National Academy Press.

———. Committee to Review Risk Management in the Department of Energy's Environmental Remediation Program. 1994b. *Building Consensus Through Risk Assessment and Management of the Department of Energy's Environmental Mediation Program*. Washington, D.C.: National Academy Press.

Neibuhr, Gustav. 1993. Black Churches' Efforts on Environmentalism Praised by Gore. *The Washington Post*, December 3, A13.

Ohio Municipal League. 1991. *Environmental Legislation: The Increasing Costs of Regulatory Compliance to the City of Columbus*. Environmental Law Review Committee Report to the Mayor and City Council of the City of Columbus. Ohio Municipal League.

———. 1992. *Ohio Metropolitan Area Cost Report for Environmental Compliance*. Ohio Municipal League.

Regulatory Planning and Review. 1993. *Executive Order 12866*, Sec. (b)(4) and Sec. (c)(D), 58 *Federal Register* 51735. Washington, D.C.: U.S. Government Printing Office.

Rivlin, Alice M. 1994. Rationalism and Redemocratization: Time for a Truce. In *Worst Things First? The Debate over Risk-Based National Environmental Priorities*, edited by Adam M. Finkel and Dominic Golding. Washington, D.C.: Resources for the Future.

Rodricks, Joseph V. 1992. *Calculated Risks: The Toxicity and Human Health Risks of Chemicals in Our Environment*. London: Cambridge University Press.

Ruckelshaus, William. 1985. Risk, Science, and Democracy. *Issues in Science and Technology* (Spring): 19–38.

Schneider, Keith. 1993. The Regulatory Thickets of Environmental Racism. *New York Times*, December 19, E5.

Silbergeld, Ellen K. 1993. Risk Assessment: The Perspective and Experience of U.S. Environmentalists. *Environmental Health Perspectives*. 101(2): 100–4.

Slovic, Paul, Nancy Kraus, and Vincent T. Covello. 1990. What Should We Know About Making Risks Comparisons? *Risk Analysis* 10(3): 389–92.

Starr, Chauncy. 1969. Social Benefit vs. Technological Risk. *Science* 165: 1,232–38.

Stavins, Robert, and Thomas Grumbly. 1993. The Greening of the Market: Making the Polluter Pay. In *Mandate for Change*, edited by Will Marshall and Martin Schram. New York: Berkeley Books.

U.S. EPA (Environmental Protection Agency). 1993. Two Faces of Risk. *EPA Journal* 19(1): 19.

———. Office of Policy Analysis. 1987. *Unfinished Business: A Comparative Assessment of Environmental Problems*. Washington, D.C.: U.S. EPA.

———. Science Advisory Board. 1990. *Reducing Risk: Setting Priorities and Strategies for Environmental Protection*. Washington, D.C.: U.S. EPA.

U.S. GAO (General Accounting Office). 1991. *Environmental Protection: Meeting Public Expectations with Limited Resources*. Washington, D.C.: U.S. GAO.

U.S. OTA (Office of Technology Assessment). 1993. *Researching Health Risks*. Washington, D.C.: U.S. Government Printing Office.

U.S. Senate. 1993. *Environmental Risk Reduction Act of 1993*. 103rd Congress, 1st session, S. 110.

5

Refining the CRA Framework

John D. Graham and James K. Hammitt

The American public has made an enduring commitment to public sector programs aimed at reducing risks to human health, safety, and the environment. This commitment is evident not just in public opinion surveys but also in the behavior of elected officials, who have passed ambitious legislation aimed at reducing risks and who have built numerous administrative agencies with far-reaching powers to influence the behaviors of business firms, workers, consumers, and the general public. The budgets appropriated to these agencies by elected officials often have fallen short of what is necessary to achieve such lofty social objectives, but risk protection agencies survived the years of "regulatory relief" under President Ronald Reagan and are now the subject of renewed interest in policy circles.

The need for risk reduction agencies such as the U.S. Environmental Protection Agency (EPA) and the Occupational Safety and Health Administration (OSHA) is widely recognized, yet serious questions have been raised in recent years about whether these agencies are directing scarce resources toward the most serious threats. A large body of evidence suggests that the nation could achieve a much larger reduction of overall risk if the same amount of resources were reallocated from less serious risks to more serious risks (Breyer 1993). A broad (if not universal) political consensus is emerging that some form of comparative risk analysis (CRA) should be used to assist Congress and

John D. Graham is professor and director at the Center for Risk Analysis, Harvard School of Public Health. James K. Hammitt is associate professor at the Department of Health Policy and Management and at the Center for Risk Analysis, Harvard School of Public Health. The authors thank Sandra Baird, Joshua Cohen, Allison Cullen, March Sadowitz, Kimberly Thompson, and participants at the RFF Workshop on Comparative Risk, held February 16, 1994, for their helpful comments.

administrative agencies in setting priorities for risk reduction (Carnegie Commission 1993).

This chapter provides some conceptual guidance to administration and congressional officials who are engaged in the task of promoting a new priority-setting process informed by CRA. The chapter begins with some clarifying comments about the case for risk-based priority setting before focusing on conceptual suggestions for refining the CRA framework. These suggestions, grounded in decision theory, point to a renewed way of implementing CRA.

THE CONSTRAINT TO RISK REDUCTION IS RESOURCES, NOT MONEY

The American people cherish good health and environmental quality. However, they are not necessarily prepared to commit to risk-reduction goals without knowing something about how much of our nation's scarce resources are likely to be devoted to the commitment. The scarcity problem has little to do with the supply of money per se, which is simply green paper (or, more essentially, electromagnetic charges in computer memory) used for exchange purposes. Scarcity arises from limitations on the quantity and quality of productive inputs such as human talent, raw materials, media time, human attention spans, and political courage.

The social objectives that compete with risk reduction for resources include such worthy activities as education, child care, violence prevention, health care, transportation, civil rights, housing, national defense, and production of various other consumer products and services. While in some cases more than one social objective can be advanced with the same allocation of scarce resources, in many cases the precious resources used in pursuit of one objective cannot be applied to another. It is the scarcity of resources combined with multiple social objectives that force Americans to make hard choices about which risks deserve high priority and which must be accorded lower priority.

The unwillingness of the public to give risk regulators unlimited access to scarce resources is increasingly evident in the fields of environmental protection and health care. Large majorities of voters in California (1991), Ohio (1992), and Massachusetts (1992) rejected ambitious ballot initiatives aimed at environmental protection, in part because of concern that the perceived reductions in risk did not justify the perceived resource commitments in a period of economic recession. The basic health care package proposed in 1994 by Hillary Rodham Clinton and her associates is a remarkably explicit illustration of resource rationing: witness the decision to authorize cervical cancer screening every three years instead of every year and breast cancer screening only every two years and only for women above age fifty (Graham 1994).

Idealists, including some prominent leaders of the public interest and environmental advocacy communities, are concerned that the growing emphasis on comparing risks is misplaced. Various technical and ethical arguments are made against risk-based priority setting, but it is important to recognize the environmental advocate's professional priorities. The advocate may believe it is more essential to fight for increases in the size of the resource pie available for risk reduction than to participate in efforts to make sure that every slice of the current pie is allocated optimally. Indeed, the process of rationing resources for risk reduction is no more attractive to environmental, labor, and consumer advocates than it is to health care professionals now struggling with the future of America's troubled health care system.

Some have argued that there are defensible reasons for hiding the reality of rationing through symbolic politics (such as mandating "safety" or "clean air" or "comprehensive health care" without regard to resource cost). We believe it is far preferable to establish priorities through an explicit process of scientific and political deliberation. Moreover, the case for CRA rests on the desire to promote not just the value of efficiency ("bang for the buck") but also the values of civic education and self-governance (Landy, Roberts, and Thomas 1990). For if governmental elites hide the reality of rationing from the citizenry, then how are citizens participating in their government in any meaningful way? Contrary to the suggestion that CRA must be a tyranny of experts, it can instead be seen as a framework that can be used to advance Jeffersonian ideals in an arena where both science and values are complex and controversial.

The public supports the new focus on comparing risks: our November 1993 survey of 1,000 U.S. citizens found that over 80% of respondents agree that "the government should use risk analysis to identify the most serious environmental problems and give them the highest priority in spending decisions" (HCRA 1994). The time has come for risk analysts to provide their best guidance on how the CRA process should be implemented, recognizing that the analyst's dream may bear only a rough resemblance to what is appropriate in a Jeffersonian democracy.

TIME FOR GUIDANCE: FINDING STANDARDS
FOR MEASURING RISK

The well-documented efforts of three successive EPA administrators—William Ruckelshaus, Lee Thomas, and William Reilly—to compare environmental risks have certainly played an important role in building the political case for setting risk-based priorities throughout the federal government. The Bush administration deserves credit for beginning discussions of the need for stronger White House leadership on CRA, but it made only limited progress on the issue. It is encouraging that the Clinton administration and the 104th Congress are now considering a variety of serious initiatives aimed at allocating risk reduction resources more efficiently, more equitably, and more explicitly.

As we move forward, it is important to remember that there is no such thing as a value-free ranking of risks. In previous comparative risk efforts directed at environmental problems, the assigned task was to rank risks according to some (implicitly defined) conception of seriousness. Although the definition of "risk" is simply "the chance of suffering harm or loss" *(American Heritage Dictionary)*, participants in risk-ranking exercises are responsible for deciding how much importance to assign to differences in probability versus differences in the magnitude of the harm, and how to account for qualitative differences in possible outcomes.

Many different risk metrics—standards of measurement of risk—can be used. In evaluating threats to human health (for example, cancer caused by chemicals released to the environment), risks are often measured by the increase in the probability of premature death for an individual, the average diminution in lifespan of those exposed, or the statistically expected number of individuals who will succumb to the hazard. (Other dimensions of the possible harm, including morbidity, can also be included in the measure.) The choice of metric can obviously affect the risk ranking: accidents will be considered relatively more important than cancer-causing chemicals if risk is measured by expected number of years of life lost rather than by the increase in mortality probability, because accidents typically kill people at younger ages than do cancers. Risks that affect consumers or other large populations will rank higher relative to those affecting smaller groups (certain types of workers, for example) if risk is measured by the expected number of premature deaths rather than by the increased mortality risk for an individual.

When we turn from human health risks to ecosystem risks, the problems multiply. In contrast to human health, no standard measures exist for risk to ecosystems or environments. Previous efforts to rank ecosystem risks have highlighted the persistence and geographic extent of consequences (U.S. EPA 1990). Even if these characteristics are accepted as a basis for comparing ecosystem risks, there is no accepted basis for comparing risks to ecosystems with those to human health. Thus, a judgment that the baseline risk (in the absence of new policies) of, say, habitat destruction is greater than the baseline risk of, say, toxic air pollution reflects a combination of scientific and value judgments. Additional efforts to quantify and rank ecosystem risks are needed.

Our suggestions in this chapter for improving comparative risk analysis are based on our understanding of how EPA and the states have used the existing analytic framework in the past. Several of these suggestions elaborate on comments that one of us previously made to Senator Daniel Patrick Moynihan with respect to improving S. 110, the risk-ranking bill Moynihan proposed in the 103rd Congress (Graham 1993). Although Moynihan's bill would have covered only EPA (primarily due to Moynihan's committee assignment), we believed then, as we do now, that it is important to recognize that environmental policymakers are already making greater use of CRA than are risk regulators at OSHA, the Food and Drug Administration (FDA), the Consumer Product Safety Commission (CPSC), and the Department of Energy (DOE). Broad-based policy from the White House and the Capitol is needed to institutionalize the CRA framework throughout the federal government.

APPLYING DECISION THEORY: FIVE SUGGESTIONS FOR REFINING THE ANALYTICAL FRAMEWORK OF CRA

The five suggestions presented below are intended to ground risk-based priority setting more securely in the canons of decision theory. This branch of management science is well suited for solving complex problems where strong contributions from both the natural and social sciences are required (Raiffa 1968). In the discussion of each suggestion, we explain how risk-ranking exercises could be improved by a more solid grounding in management science. Our suggestions for refining the CRA framework, which forms the basis for risk ranking, are:
- Rank both risk-reduction options and baseline risks.
- Rank according to the expected value of risk reduction.
- Use uncertainty as an information source.
- Rank according to both competing risks and target risks.
- Rank according to both resources costs and savings.

Rank Both Risk Reduction Options and Baseline Risks

In the future, the question should be reformulated as a ranking of risk-reduction options as well as a ranking of risks. Ranking risk-reduction options is admittedly a more complex task. It asks experts to both forecast how much risk might be reduced under various risk management alternatives and rank options according to risk-reduction potential.

From both technical and practical points of view, there are strong reasons for ranking risk-reduction options as well as ranking baseline risks.

Concentrating resources on the largest risk may be unwise if little can be done to reduce it. If a 90% reduction can be made against the third-largest risk (C) and only a 10% reduction can be made against the largest risk (A), the overall risk reduction may be greater if extra resources are applied to risk C. Even if the risk from collision with an asteroid exceeds that of cancer from the electromagnetic field around an electric blanket, it may be better to direct attention to electric blankets. This insight will be revealed by a ranking of risk-reduction options but may not be obvious to a decisionmaker who is simply provided a ranking of baseline risks.

Similarly, looking at the problem "risk by risk" conceals the promise of solutions that can reduce more than one risk. Even if risk A is the worst risk, it may be that an innovative policy option can reduce or even eliminate both the second- and third-worst risks (B and C). The resulting risk reduction may be more than can be achieved by even the best policy against risk A. For example, even if we assume that the risk due to enhanced skin cancer from chlorofluorocarbon-induced ozone depletion exceeds the risks both to farmworkers and to consumers from using synthetic pesticides on food crops, shifting to organic farming may reduce more risk than banning chlorofluorocarbons would, because organic farming affects both farmworker and consumer risks.

Ranking decision options can also minimize disconcerting ambiguities about how risks should be aggregated for purposes of ranking baseline risks. For example, should the ubiquitous "criteria" air pollutants and "hazardous" air pollutants be ranked separately or ranked under the general heading of "air pollution"? This is a salient yet intractable question for risk rankers but poses no ambiguity when decision options are ranked. If an option reduces both types of pollution, the resulting risk reductions are summed (absent interactions). An option that only reduces hazardous air pollution is given no credit for reducing criteria air pollution.

This seemingly picky point about risk aggregation indeed has been a source of considerable consternation among participants in EPA-sponsored efforts to rank baseline risks. After all, there is no right answer to the question of how risks should be aggregated and listed for purposes of ranking. Comparable ambiguities arise when risk-reduction options are ranked (for instance, how broad or aggregate the range of decision options should be). However, in this case there are decision-analytic procedures for determining the optimal portfolio of risk-reduction options.

Those readers who regard the points just made as elementary may find it useful to read the reports *Unfinished Business* (U.S. EPA 1987) and *Reducing Risks* (U.S. EPA 1990), considered by many to be the most serious analytical efforts at large-scale CRA. These reports give primary emphasis to risk ranking while containing little serious analysis of the relative promise of various risk reduction options. We as a nation can do better in the future.

Rank According to the Expected Value of Risk Reduction

Faced with uncertain scientific knowledge about risk, decision options should be ranked according to the *expected reduction* in risk, although decision options that reduce catastrophic possibilities deserve explicit upgrades in the risk-ranking process.

When uncertain estimates of risk are compared, it is important that the relative degrees of scientific uncertainty be handled in a careful and consistent fashion. For example, a comparison of decision options A and B can be misleading if the risk-reduction estimates for A are made optimistically (that is, assuming worst-case estimates of baseline risk and high degrees of risk-reduction effectiveness) while the risk-reduction estimates for B are made pessimistically (that is, assuming lower-bound estimates of baseline risk and lower-bound estimates of effectiveness).

Since risk regulators will face many risk reduction decisions over time, it would be useful to adopt a priority-setting strategy that maximizes the cumulative amount of risk-reduction achieved. A procedure to attain this objective is to:

1. Estimate the probabilities of various degrees of risk reduction for each decision option using the best available science.
2. Compute the expected amount of risk reduction for each option (where various risk reduction possibilities are weighted by their probability of occurrence and then summed).
3. Rank decision options according to the "mean" estimate of risk reduction (that is, the expected value).

The attractiveness of this procedure rests on the law of large numbers and the properties of the arithmetic average as a summary statistic (Zeckhauser and Viscusi 1991). If uncertain risk-reduction opportunities are ranked using any criterion other than the expected value, the maximum possible degree of risk reduction will not be achieved in the long run.

Our mentor Professor Howard Raiffa has commented that the failure to consider expected values of risk in priority setting may result in the "statistical murder" of citizens and the environment. Those who advocate ranking of risks according to "conservative" or "worst-case" estimates of risk will not achieve their stated goal of providing the maximum degree of protection to public health and the environment. Those who advocate ranking risks according to the "most likely" or "most probable" estimate of risk often do not recognize the distinction between the mean (or "expected value") of a probability distribution and other summary statistics, such as the mode or the median. The expected value is sometimes strongly influenced by best- and worst-case possibilities (even if they are unlikely to occur) while the median and modal values assign no weight to unlikely possibilities. In some cases, it has been shown that the mean of a distribution of possible cancer potency values exceeds not only the median and mode of the estimate, but even the upper 95th percentile of the distribution (Hattis 1990).

Decision theorists have demonstrated exceptions to the guideline that priorities should be based on the expected value of risk reduction. Just as families will be rationally averse to seemingly attractive gambles where the downside losses approach financial ruin for the family, the nation or planet should be especially averse to health, safety, and environmental risks that have potentially catastrophic outcomes, such as some of the more dangerous outcomes of global climate change. Moreover, risks that are correlated (whether geographically or temporally) among individuals or populations may be viewed as more serious than similar but uncorrelated risks, because when risks are uncorrelated, those who do not suffer the harm may be able to assist those who do (this, incidentally, is the principle underlying the insurance industry). The use of geographic scope and permanence as partial indicators of the severity of ecosystem harm is consistent with concern for correlation, since a widespread, irreversible harm (such as the extinction of a species) is one that is correlated across individuals.

In decision theory, the concept of "risk aversion" has been introduced to reflect rational departures from the use of expected values when making decisions under uncertainty. Consider for example the risk of all-out nuclear war (or even destruction of a major city): this should be considered a more serious risk than would be revealed by the simple product of the probability of the war multiplied by the resultant loss of human life and ecosystem damage. The extent to which rankings should depart from those suggested by the expected risk depends in part on the risk metric and the associated degree of risk aversion.

Although aversion to comparatively modest disasters, in which tens or hundreds of people may die, is a legitimate public value, we caution analysts not to overestimate the importance of risk aversion. Each departure from expected-value decisionmaking reduces the total number of lives or life years that may be saved. We suspect that resources are disproportionately allocated to preventing disasters because of public aversion to any concentrated loss of life, the newsworthiness of such events, and desire of officials to avoid being held to blame when disasters occur. CRA may show that substantial gains in lifesaving would be achievable if resources were partially reallocated from these high-visibility risks to more commonplace risks that strike victims individually (for example, from airline safety to automobile safety). Moreover, when those who perish in a disaster are members of the same families and communities, the total grief and suffering in some sense may be less than when many dispersed families and communities lose loved ones.

Use Uncertainty as an Information Source

If conducting additional research prior to taking action is a viable decision option, or if actions can be revised based on early results, then risk-ranking exercises should also categorize decision options according to degree of scientific uncertainty about risk reduction and resource cost and about the possibility of obtaining useful information about them.

If the opportunity to learn more about risks before acting is available (or if actions can be revised in the light of new information), actions about which there is greater uncertainty become relatively more attractive. The potential benefits from learning about a risk-reduction option by studying it or trying it (perhaps on a limited basis) depend on two factors. First, the risk reduction or resource costs of the option must be sufficiently uncertain that the desirability of undertaking the option is unclear. Second, feasible research approaches must offer the ability to reduce existing uncertainty enough so that a decision other than the one that would be made without additional information may be selected.

Consider a choice between two drinking-water purification systems. One (chlorination) is known to work effectively but to produce a small risk of cancer from its by-products. For illustration, assume that this risk is known to be either nine or eleven premature deaths per million lifetimes, where each risk has 50% probability. The (hypothetical) alternative is known to be free of adverse side effects but may or may not work effectively. If it does work effectively, it poses no risk; if not, fifty premature deaths per million lifetimes are expected due to system failure. Assume the chance the new system works effectively is 50%.

If it is not possible to learn more about the hypothetical system's efficacy, chlorination is the better alternative, because the expected risk of chlorination, ten premature deaths per million lifetimes, is less than that of the alternative, twenty-five premature deaths. But suppose it is feasible to undertake research that will reveal the exact risk from either chlorination or the alternative. If chlorination is studied, it will turn out to pose a risk of nine or eleven premature deaths per million lifetimes. In either event, however, chlorination will remain preferred to the alternative system. Because the research cannot provide an estimate that is sufficiently different from the expected value of the risk, it cannot provide information that changes the decision that would otherwise be appropriate. On the other hand, if the alternative is studied, there is a 50% chance of discovering that it is perfectly safe and effective, and hence superior to chlorination. Research on the alternative system therefore has a 50% chance of yielding information that will lead to a change of practice and a reduction in risk.

In this hypothetical situation, a complete analysis of research options must also consider the costs of learning about the new system, the probability that research will provide relevant knowledge, the effects of delaying a decision, and other factors. Greater uncertainty about the risk of a decision option can provide an opportunity for learning, and potentially for greater risk reduction, than if the risk or control strategy were known with certainty. "Value-of-information" methods have been developed by decision analysts to address this kind of complication and have been applied to problems ranging from air pollution control (Finkel and Evans 1987) to food safety (Hammitt and Cave 1991).

Rank by Both Competing Risks and Target Risks

Risk-reduction options should be ranked in terms of *net* risk reduction, taking into account competing risks as well as target risks.

Most CRAs simply rank various "target risks" whose baseline levels (prior to policy action) are of concern to policymakers. A limitation of this approach is that it does not inform decisionmakers of various risks that are caused by efforts to reduce a target risk. Here we define a "competing risk" as a health, safety, or environmental danger that is caused by a policy option whose ostensible purpose is to reduce a target risk (Graham and Wiener 1995).

Competing risks, while ubiquitous, are frequently neglected in risk management. Some options that reduce air pollution create more wastes that must be incinerated or disposed in landfills, both of which present competing risks to the public and the environment. Making cars smaller and more fuel efficient reduces various environmental risks but may increase the danger to motorists when crashes occur. Banning one solvent, a known human carcinogen, creates a competing risk if workers will be exposed to a substitute solvent that is a suspected carcinogen. Long-term postmenopausal estrogen therapy reduces the risks of osteoporosis and coronary heart disease but may increase a woman's risk of breast cancer.

The technically appealing approach to this problem of competing risks is to rank options in terms of net risk reduction instead of simply comparing opportunities to reduce target risks. We acknowledge the increased complexity that is raised by the phenomenon of competing risks. The analyst must not simply forecast the change in any target risk but also identify and quantify any competing risks and compare them to the change in the target risks.

One is tempted to resist this complication on the grounds that the target and competing risks may be qualitatively different and hence difficult to add and subtract on the same scale. While the challenge of developing a universal risk metric is formidable, that is precisely the analytical challenge faced by risk rankers. If it is feasible to make rough comparisons of highly diverse target risks for purposes of ranking, then it should be possible to make rough comparisons of target and competing risks. After all, a target risk for one policy option (such as groundwater contamination to be addressed by pumping and treating) may prove to be a competing risk for another policy option (for instance, making greater use of potentially leaky landfills instead of incineration).

We sympathize with the position that competing risks should be addressed only after the most serious target risks have been identified as priorities and specific risk-reduction options developed for them. If we analyze too much at the priority-setting stage, we may suffer paralysis by analysis. The drawback of censoring consideration of competing risks, however, is that priorities will be misordered. A large risk that can only be reduced through options that create onerous competing risks should not necessarily be given high priority. Much of medicine falls in this class—the treatment can be harmful, but failure to treat is usually worse (such as AZT treatments for AIDS, chemotherapy for cancer). A target risk of only modest size may actually deserve high priority if it can be reduced substantially without producing significant competing risks.

In summary, there are sound reasons for ranking options according to net risk reduction rather than simply ranking baseline risks or ranking options in terms of the degree of reduction in the target risk.

Rank by Resource Costs and Savings (Cost-Effectiveness Analysis)

Risk-reduction options should be ranked according to resource cost-effectiveness criteria; that is, the ratio of the expected net risk reduction to the expected resource cost, taking into account any resource savings resulting from the option.

It is tempting to ignore resource costs in priority setting because:
- the magnitude of resource costs associated with various decision options is often uncertain and difficult to estimate, and
- the explicit accounting of risks and resource costs may offend those who object to the practice of monetary benefit-cost analysis.

While we as researchers are wary of analytical efforts aimed at reducing all consequences to a dollar equivalent, there are strong reasons for considering the resource implications of alternative risk-reduction options.

Even if a risk-reduction option promises a large amount of risk reduction, it may not justify a high priority if the option would place greater demand on scarce resources than the nation currently is willing to commit. For example, a national transition from the internal combustion engine to electric cars would require an enormous expenditure of resources, at least during the transition period. Faced with this reality, the Clinton administration and the motor vehicle manufacturing industry recently opted for an accelerated program of research and development aimed at reducing the resource costs and increasing the quality of new vehicle technologies, thereby making them more feasible in the marketplace.

A useful way to compare benefits with resource costs that is widely employed in medicine is to rank risk-reduction options according to the expected ratio of net risk reduction to net resource costs (Weinstein 1990). (If a decision option will expend some resources while saving others, the analyst is asked to compute the net change in resource use, usually in monetary units). Decision options are then ranked from those with the highest ratio to the lowest ratio of risk reduction per unit of resource input. This strategy, often called cost-effectiveness analysis, does not require the comprehensive monetization that is required by benefit-cost analysis, although it forces decisionmakers to make judgments about which ratios are promising enough to justify investments in risk reduction.

At the Harvard Center for Risk Analysis, we are assembling a database of cost-effectiveness ratios for all decision options that reduce the probability of premature death. We currently have information on the cost-effectiveness of over 500 decision options in medicine, consumer product safety, environmental and occupational health, and transportation safety. We have found that, in the case of lifesaving investments, many highly cost-effective opportunities are not implemented while more extravagant investments with limited lifesaving promise are made without controversy. For example, shoulder belts have been added to the rear seats of automobiles at a cost of $5 million per year of life saved while government and manufacturers have been slow to add airbags to the front-right seat of passenger cars at marginal costs of less than $1 million per year of life saved (Bell 1994).

In the future, we would like to see each regulatory agency publish such cost-effectiveness figures for all of the risks within the agency's jurisdiction. The agency would then be required to justify situations where programs with especially low returns were adopted and programs with much higher returns were rejected.

When considering the resources consumed and saved by risk reduction options, it is critical to consider all resources that are consumed or saved, regardless of whether they are reflected in the federal budget. At agencies such as EPA, the resource cost to EPA is often small compared to the total resource cost of risk reduction because the bulk of the payments for resource use are shifted to states, localities, and the private sector. The political movement to reconsider "unfunded mandates" by the federal government is not simply a movement for greater federal subsidies to state and local governments, but also a movement for careful scrutiny of the total resource costs (and savings) associated with risk reduction.

CONSIDERING BOTH RISK EQUITY AND RISK EFFICIENCY

Decision theorists have long recognized that good decisions may reflect various principles of fairness or equity as well as efficiency. Resource allocation decisions for risk reduction should also be made with equity considerations in mind. A good illustration for reflection is the case currently being made by environmental justice advocates that the risks of pollution are incurred disproportionately by low-income and minority populations.

Suppose three risks, A, B, and C, are ranked equally according to the expected number of lives lost because each is expected to kill roughly the same number of people. If C affects primarily low-income and minority populations (while A and B are distributed evenly by income and racial groups), then a strong equity argument can be made that risk C deserves higher priority than risks A and B. Some equity advocates might reason from John Rawls' difference principle that priority should be given to actions that reduce risks to the most disadvantaged, regardless of the expected number of lives lost among nondisadvantaged citizens. Alternatively, Rawls' principle might be interpreted to suggest assigning priority to risk reductions that benefit those who face the largest individual risks from all causes (environmental and others).

Equity arguments are important and more subtle or complicated than they appear at first glance. They therefore need to be analyzed carefully because each departure from expected-value decisionmaking will reduce the overall amount of risk reduction achieved for a given quantity of resources expended. Redistribution of risk reductions, as of other goods within society, may be achieved more efficiently by programs that transfer general resources like income (with which individuals can purchase risk reductions and other goods) than by altering health and environmental programs to explicitly promote equity. Perversely, the notion of assigning highest priority to risks that affect individuals who already face large risks from other causes may be particularly inefficient, since even the elimination of one risk may only marginally extend the lifespan of an individual because one of the competing risks will prove to be lethal.

If equity is to be analyzed properly, the precise notion of equity must be well-defined and risk analysts must be instructed to collect data and report estimates of risk that facilitate equity determinations. While some advocates of environmental justice are wary of risk analysis (in part because of the way it may have been used by some decisionmakers to site hazardous facilities in minority communities), we see comparative risk analysis as a promising ally for those concerned that insufficient resources have been dedicated to improving the welfare of low-income and minority populations. We suspect that many of the risks in America that would score highly in risk-ranking exercises are indeed ones that strike the poor and disadvantaged citizenry with disproportionate frequency (such as childhood lead poisoning and urban air pollution). By making data and assumptions about the probability and consequences of alternative risks more important in decisionmaking, comparative-risk analysis may advance equity by impeding those who would otherwise manage to redirect resources only toward risks that affect the wealthy and politically powerful.

USING THE NEW FRAMEWORK:
BUILDING INSTITUTIONAL CAPACITIES

Although the intellectual challenge of comparative risk analysis is enormous and the supply of talented analysts limited, it is our opinion that the greatest obstacle to progress today is institutional rather than intellectual. The federal government possesses few institutions with the credibility and mix of skills to perform such analyses. Ironically, EPA—the federal agency that has been most criticized for a failure to base priorities on risk—is probably the federal institution best equipped to make important strides in the right direction. Since institutional change of any sort is difficult, we believe it will take strong signals from the White House and Capitol Hill to move the comparative risk agenda.

The Executive Office of the President possesses limited expertise on the technical challenges of risk analysis. The Office of Management and Budget and the Council of Economic Advisers have significant economic expertise but only limited access to expertise from the hard sciences, engineering, mathematics, and statistics. The Clinton administration's new Office of Environmental Policy has good policy and political skills but is short on scientific talent. While the Office of Science and Technology Policy (OSTP) would seem to be the logical home for risk analysts in the White House, OSTP has taken only sporadic interest in risk issues throughout its history. Any serious administration move to promote comparative risk analysis must address the need for risk analysis expertise in the Executive Office of the President.

Congress plays a crucial role in risk-regulation policy but is also poorly organized and staffed to promote the comparative risk perspective. Both the personal and committee staffs continue to be dominated by attorneys, although some scientists have been hired in recent years. The General Accounting Office, the Congressional Research Service, and the Office of Technology Assessment have commented periodically on risk analysis issues, but none of these organizations has a broad-based technical capability in this area. In the long run, fixing the misprioritization of risks in national policy may require a new organizational structure in Congress such as a joint House-Senate committee on comparative risk topics. Such a body would be responsible for looking beyond the jurisdictions of parochial committees and subcommittees and for pinpointing the most promising risk reduction opportunities.

The administrative and cabinet agencies—such as EPA, FDA, OSHA, DOE, CPSC—themselves have a significant amount of risk analysis expertise, but most of it is confined to a particular kind of risk (such as EPA's experts on the risks of toxic pollution). The suggestion of a centralized risk analysis office within EPA is attractive because it would facilitate a synoptic view of environmental risks that is necessary for good resource-allocation decisions. Few analysts in agencies have had the opportunity to move around the federal government and develop broad-based expertise in the process of estimating risks and comparing them for the purposes of priority setting. Justice Stephen Breyer has made a constructive proposal to cultivate a cadre of career experts in risk analysis who would advise the president based on a track record of work in numerous agencies (Breyer 1993). This idea deserves serious consideration.

Having criticized the institutional capacity of the federal government, we must be honest and acknowledge that the university community has been slow to recognize the crucial role of risk analysis in the training of scientists, lawyers, business people, and public managers. At the Harvard Center for Risk Analysis, we are determined to make a long-term contribution to the training of professionals who will ask the right questions about risk. A more important step, though, may be for elementary and secondary schools to do a better job educating students in basic skills of math, science, and economics. Without both congressional and presidential leadership, it is unlikely that the campaign for risk literacy will succeed. We are therefore encouraged by the current efforts underway in Congress and the Clinton administration.

An analyst's fantasy about CRA might involve a monumental commitment to data collection, analytic modeling, and conceptual theorizing. Although we recognize that our suggestions would require extraordinary analytical effort to implement fully, we believe that CRA can be substantially improved by using the tools of management science to develop a clear and consistent conceptual focus. We cannot let contemplation of the ideal analysis prevent attainment of the good policy. Simplifications required by practicality will be least harmful if they are analytic approximations that are explicitly chosen while keeping the underlying decision-analytic framework firmly in mind. Our suggestions for expanding the analytical framework of CRA represent a step in that direction.

REFERENCES

Bell, Nicole. 1994. Personal communication to the authors.

Breyer, Stephen. 1993. *Breaking the Vicious Circle: Toward Effective Risk Regulation.* Cambridge, Massachusetts: Harvard University Press.

Carnegie Commission on Science, Technology, and Government. 1993. *Risk and the Environment: Improving Regulatory Decision Making.* New York: Carnegie Corporation.

Finkel, Adam M., and John S. Evans. 1987. Evaluating the Benefits of Uncertainty Reduction in Environmental Health Risk Management. *Journal of the Air Pollution Control Association* 37: 1164–71.

Graham, John D. 1993. Letter to Senator Daniel Patrick Moynihan, 4 August.

———. 1994. The Role of Risk Analysis in Environmental Protection. Testimony given at a public hearing before the Committee on Government Operations, U.S. House of Representatives, Washington, D.C., 1 February.

Graham, John D., and Jonathan B. Wiener. 1995. *Risk Roulette: Confronting Tradeoffs in Health, Safety, and Environmental Decisions.* Cambridge, Massachusetts: Harvard University Press.

Hammitt, James K., and Jonathan A. K. Cave. 1991. *Research Planning for Food Safety: A Value-of-Information Approach.* Santa Monica: Rand Corporation.

Hattis, Dale. 1990. Letter to the Editor. *Risk Analysis.*

HCRA (Harvard Center for Risk Analysis). 1994. Does the Public Support Risk Analysis? *Risk in Perspective* 2(1).

Landy, Marc K., Marc J. Roberts, and Stephen R. Thomas. 1990. *EPA: Asking the Wrong Questions.* New York: Oxford University Press.

Raiffa, Howard. 1968. *Decision Analysis: Introductory Lectures on Choices Under Uncertainty.* Reading, Massachusetts: Addison-Wesley.

U.S. EPA. (Environmental Protection Agency). Office of Policy Analysis. 1987. *Unfinished Business: A Comparative Assessment of Environmental Problems.* Washington, D.C.: U.S. EPA.

———. Science Advisory Board. 1990. *Reducing Risks: Setting Priorities and Strategies for Environmental Protection.* Washington, D.C.: U.S. EPA.

Weinstein, Milton C. 1990. Principles of Cost-Effective Resource Allocation in Health Care Organizations. *International Journal of Technology Assessment in Health Care* 6: 93–103.

Zeckhauser, Richard J., and W. Kip Viscusi. 1991. Risk Within Reason. *Science* 238: 559–64.

6

A Proposal for Ranking Risk within Federal Agencies

M. Granger Morgan, Baruch Fischhoff, Lester Lave, and Paul Fischbeck

Scarce time and resources prevent individuals and society from doing everything they might to reduce risks to health, safety, and the environment. Risk management priorities are often set through somewhat chaotic social and political processes. A systematic ranking of risks can help federal risk managers to evaluate whether they are allocating their attention and society's resources in a sensible fashion.

The U.S. Environmental Protection Agency (EPA) has already completed two major risk-ranking exercises. It also has assisted state and regional organizations in risk rankings. A goal of many of these state and local ranking exercises has been to mobilize public opinion and local interests in support of improved risk management. In contrast, for federal risk management agencies, a more analytical objective is appropriate. If, with expert and public input, federal risk management agencies systematically rank the risks for which they are responsible, the process of ranking should help them to think more clearly about the choices they face. The results of the ranking should prove useful in both short-term agency and long-term public and congressional evaluation of risk management priorities.

This chapter describes a procedure that federal risk management agencies can adopt to perform such ranking. It calls for creating an interagency risk-ranking task force to coordinate the effort. Under guidance from this task force, agencies define and categorize the risks

The authors are members of the faculty of the Department of Engineering and Public Policy at Carnegie Mellon University. They were assisted in this work by graduate students Stephanie Byram, Karen Jenni, Garrick Louis, Sandra McBride, Laura Painton, Stuart Siegel, and Ned Welch.

that they will rank. In parallel, the task force identifies the attributes that all agencies will use in evaluating "their" risks. Using appropriate experts, each agency develops concise quantitative evaluations of each risk in terms of these attributes, together with a qualitative description. Careful attention is devoted to describing uncertainty. The results for each risk are published on standardized summary sheets.

Ranking for each agency is performed by four groups: a team of federal risk managers drawn primarily from the agency, but including a few representatives from other agencies; two groups of lay people selected through a quasi-random process; and a group of state and local risk managers. These groups receive staff support but are largely responsible for their own deliberations.

The results of a ranking exercise can be very sensitive to the procedures employed. In order to minimize procedural dependency, each ranking group will employ two quite different procedures: one reductionist and analytic, the other holistic and impressionistic. Groups will use the (possibly different) results of the two procedures as an aid in developing and refining their informed preferences. Attention is directed at getting risks ranked in the right "ball park." When large uncertainties preclude more precise ranking, they are not pursued.

At the end of the exercise, the four groups will come together to examine the extent to which they share similar views. When the groups' views can be reconciled, this is done. When reconciliation is not possible, the reasons are carefully explained. In addition to ranks, the final report from the process provides a reasonably rich qualitative and quantitative description of the deliberations and the results.

To better understand this proposed procedure, this chapter includes considerable contextual detail. The chapter's sections cover the following topics in the order noted:

- motivations for, and difficulties involved in performing, risk ranking
- the recent historical background from which the general issues of risk-ranking exercises have emerged
- the specific challenges facing risk ranking in the federal government and the design philosophy needed to cope with them
- a translation of this design philosophy into a design strategy
- the proposed procedure itself
- the steps needed to implement and evaluate this procedure

WHY PERFORM RISK RANKING?

Risks are everywhere. How do we decide which ones are worthy of our attention? In an ideal world, we would regularly review our priorities.

In reality, though, such systematic reviews of risk are as rare as systematic reviews of how we spend our attention, money, or emotions. Time is one obvious constraint on such reviews. However, even with all the time in the world, there would still be daunting obstacles. Risks are so diverse that it is hard to compile either the list of threats or the list of control strategies. Often, we have inadequate estimates of the sizes of the risks, the chances for control, or the costs of amelioration.

Even if we had all the figures, we would face trade-offs in ranking risks that confront us with difficult ethical, social, and emotional choices. What we do about risks defines us as people and as citizens. It shows what we value and what we accept as our personal (or social) responsibility. Any risk that we neglect can come back to haunt us. Any risk that we face (or explicitly neglect) raises uncomfortable concerns. Any risk that we place on our own agenda may be taken off the plates of others, such as those who create it (and benefit from it).

As a result, both individually and as a society, we usually just muddle along. We have evolved a set of practices to deal with risks. Some practices we have adopted through deliberate rational analysis. Others were imposed upon us by political or other imperatives. Still others have evolved in socioeconomic contexts that paid little or no attention to safety. Every once in a while, something happens that calls these practices into question. It may make us wonder whether we are needlessly investing in risk control or recklessly leaving ourselves exposed. On the positive side, these confrontations offer sequenced opportunities to rethink our priorities. They may put the same risk on many people's agendas, creating opportunities for discussion. These episodes may be accompanied by ready access to relevant information. They may be prompted by the discovery of new information. They may facilitate collective action or legitimate changes in long-standing behaviors. They can force us to think about unpleasant topics that we would otherwise ignore.

Having our societal agenda set in this way, however, has its limitations. Many of these limitations are fairly obvious if one reflects on the process, but much less so when one is mired in it. One limit is that the nomination process that determines which risks we address can have little to do with the magnitude of the risks involved, the usefulness of any new information, or the opportunities to take action. Scientists and analysts can seize center stage with studies that mean a lot to them personally, but that add little to our overall understanding of a risk or that greatly clarify a risk of little practical importance. News media often retell familiar stories, and train their audiences to expect them, while neglecting more serious risks. The public focus on a risk may create unwarranted pressure to do something about it.

When public opinion is mobilized around a topic, group processes can take on a life of their own—especially when institutions are perceived as responding callously or exploitatively. Those processes can generate their own cues as to the magnitude of the risk involved: "They're so high-handed, they must be hiding something." "They're so hysterical, how can you take their concerns seriously?" Group processes can create political and psychological stakes that come to dominate any risk concerns. Although the debate may be conducted in terms of risk, often what each side really cares about is not losing to those on the other side, who are often regarded as irresponsible (Glickman and Gough 1990; Krimsky and Plough 1988; NRC 1989).

Further complications arise when the initiating event conveys an inaccurate picture of its focal risk (or prevention strategy). For example, it is common to lack explicit quantitative estimates of the magnitude of a risk or a concise summary of the quality of the underlying science. Indeed, uncertainties might not be admitted at all (Funtowicz and Ravetz 1990; Morgan and Henrion 1991; Shlyakhter and Kammen 1993). Under these circumstances, citizens and, to some extent, even risk managers must divine the size and certainty of the risk from the indirect cues given by the initiating event. A common, reasonable, but imperfect inference is that people would not raise an issue if it were not important and they did not know something about it. The fallible converse of that inference holds that important risks get reported expeditiously, thanks in part to the large number of people—journalists, activists, scientists, and so forth—generally regarded as looking for trouble. However, the anarchy of many risk-nominating processes can also advance the cause of small, poorly understood risks and risk situations.

When experts do attempt to speak clearly about risks, there may be significant barriers to being understood. They may speak in unfamiliar jargon, using uncommonly large or small numbers expressed in strange units. They may make unwarranted assumptions regarding how much people know about the processes, principles, and controversies of science. As a result, they may unwittingly talk down to their audience or over its head. Incomprehensible messages might be ignored or misinterpreted. Such miscommunication can be inadvertent or deliberate (if there are incentives to make a risk seem large or small). In either case, a poor communication process will aggravate whatever "natural" misunderstandings people have about risks. Further problems arise when misconceptions are held with undue confidence, or accurate beliefs are given too little credence. The research literature documents a variety of such general and specific problems in lay risk perceptions and expert risk communications. It also has various bits of advice on how to alleviate problems, substantiated by varying amounts of empirical evidence (Fischhoff, Bostrom, and Quadrel 1993; Gilovich 1993; Kahneman, Slovic, and Tversky 1982; Morgan and others 1992; Slovic 1987).

Analysis for the Common Good: A Role for Government

Resolving such standoffs between the public and risk professionals is where the government can come in. Through its agencies, government has the resources, expertise, and responsibility for mustering the evidence on issues of public concern. Whatever complaints people may have about government, it still has unique credibility for conducting analyses with the common good in mind. However, government's job is a hard one. It faces the same obstacles as do individuals, only written very large and publicly.

Initially, a government agency must compile a comprehensive list of potentially hazardous situations, including those that are, for whatever reason, commonly ignored. It must summarize available scientific evidence, with adequate representation of uncertainties. For the sake of comparisons, it must reduce all risk situations to some common metric.

Each of these tasks is complicated by conventional reporting practices. Scientists and analysts may not report any numbers regarding the magnitudes of risks or uncertainties. If they do, then they may use measures of statistical significance rather than ones of practical significance (that is, answering the question of "is there a difference?" not "does the difference matter?"). They may report their results in isolation from those of their predecessors. Their systematic summary may just count the studies with and without an effect, weighting them equally despite large differences in the definitiveness of their results or the precise question being studied. Their report of uncertainty may be limited to the variability in measurements, neglecting their own concerns about the quality of the measurements themselves and the defensibility of the auxiliary hypotheses needed to interpret the data. Whatever summary measures are provided might be expressed in incommensurable units (such as changes in probability of premature death, changes in life expectancy).

Although government scientists and analysts, unlike the public whom they hope to serve, may have desks, computers, and budgets, they cannot have all the resources needed to perform all the secondary analyses involved in working up the data. As a result, they have to perform judicious compromises. For example, government analysts may choose to reject studies that fail some reporting standard, or to use some ad hoc combination rule, or to elicit summary probability distributions from some experts. Any exercise of judgment evokes some disagreement from within the scientific community. Establishing that range of opinion is an important part of characterizing risks; however, it can be hard to live with disagreements.

Value Judgments and Risk Ranking

Even with unlimited budgets, government analysts could not solve the risk-ranking problem unambiguously. Each step of the process involves value judgments. It is a question of ethics, not science, to determine which potentially hazardous situations can even be considered, how "risk" is to be measured, how the different dimensions of risk should be weighted, and how uncertainty should be treated (Crouch and Wilson 1981; Fischhoff, Watson, and Hope 1984). For example, measuring "risk" in terms of lost life expectancy will give greater weight to deaths among the young (who forfeit more days of life) than will measuring "risk" in terms of increased probability of premature death (which is indifferent to who dies).

Different resolutions of these values-related issues will lead to different rankings of risks and of risk reduction strategies. As a result, there will be differences in whose welfare is protected and whose risk-producing activities are subject to scrutiny. Often, government analysts' legal mandate provides no authority to make such choices. In those situations where the role of values in scientific analysis is recognized, there still may be no guidance on which choices should be made. Even if there is (or if there were) guidance, the choices that it dictates might lack legitimacy in the eyes of those whose fates are affected. For example, it is one thing to require benefit-cost analyses; it may be quite another to convince people of the legitimacy of value-of-a-life calculations.

Under these circumstances, government analysts can respond to such values-related issues in risk ranking in three general ways. One response is to make as few assumptions as possible, then report the results of their work in something approaching raw form. That would mean assembling the risk data (a considerable feat in itself), but not digesting it. The work product might be a set of weakly comparable estimates, expressed in different terms and with different degrees of credibility. It would be accompanied by frank discussions of the sources, assumptions, and limitations of its results. However, any integration would be left to the consumers of the data.

A second response strategy is to integrate the evidence in several different ways, reflecting alternative value systems. Doing so would not prejudge which values are appropriate (at least among those that are considered). Rather, the analysts would run the numbers for potential consumers of its results. That would let them focus on the value issues, relieved of the computational load imposed by raw data. It would allow them to make their own choice among alternative value systems. It would help to identify the extent and the sources of differences in rankings.

A third response strategy is to work with consumers of the results to develop suitable summaries of the risk data. A hands-off version of this strategy is to elicit values from consumers, then provide them with the rankings that follow from these values. The success of this strategy depends, in part, on consumers' ability to determine their values in the abstract. Often that means grappling with difficult ethical questions conveyed in an unfamiliar format and producing public statements on them. It is then up to the analysts to carry the consumers along, so that they can see the resultant rankings as expressing the values that they have provided.

A more hands-on version of this strategy involves an iterative relationship with the consumers. At each stage of the interaction, the analysts would produce partial workups of the data, incorporating whatever values emerge at the time. Over time, with the opportunity to view the task from multiple perspectives, consumers might converge on a single set of values and rankings. Or, they might recognize a range of uncertainty in what they want. Within that range, they might be particularly willing to make deals or to let policymakers assume responsibility for choices.

Thus, each of these approaches builds on its predecessor. Each, in turn, takes a successively more activist role. For example, the alternative rankings of the second approach are likely candidates for those that will be needed in the third approach. Conversely, experience drawn from interacting with consumers of data should inform the creation of alternative rankings.

Practical government involves the possible, guided by a vision of what is desirable. The strategy and processes outlined in the following sections are intended to be practical and achievable. They are intended as analytically sensible ways to help participants make sense of these complex issues. The lessons learned during the process of ranking may prove as important to improving federal risk management as the rankings themselves.

Further, while ranks may have some positive short-term impacts on federal risk management, their principal value will almost certainly be over the long term. For example, EPA's *Unfinished Business* (U.S. EPA 1987) appears to have played an important role in stimulating a reassessment of our approach to Superfund toxic waste sites. If other ranking exercises suggest that other risks are receiving too much or too little attention and if the exercises stimulate similar public, administrative, and legislative review, the process will have made a valuable contribution to improving the quality, equity, and efficiency of our risk management.

Improving federal risk management will require both heart and head. It means getting the facts right and getting people behind them. It means getting scientists to speak each other's languages and to admit to their own failings. It means getting citizens to face the tragic trade-offs required by risk management. It means getting officials to acknowledge past mistakes in the resources allocated to particular actions or programs. It means undertaking intellectually challenging analytical problems.

Such a balancing of perspectives and a broadening of accountability won't be easy. It will require carefully orchestrating both the paperwork and the human process. The stakes are very high. Perhaps the current state-level experiments, described in Chapter 3 by Richard Minard, are a sign that we are willing to cut one another enough slack to make some progress.[1]

DESIGN PHILOSOPHY

A credible risk-ranking strategy must be designed to be equally sensitive to these criteria:

- the science of risk and the difficulties in identifying, understanding, and commmunicating this science to the public;
- the values of the public, especially the limitations of using surveys to reflect those values; and
- the institutional setting within which risks are managed.

Each of these primary design philosphy critera is detailed in this section.

Specifying and Understanding the Science for Risk Ranking

Almost any risk could, in principle, come within the purview of the federal government. Even were risk ranking restricted to the reallocation of resources among currently regulated risks, an extraordinary variety of causes and effects might have to be considered. No individual or discipline could presume to understand more than a fraction of the relevant science. As a result, a wide net will have to be cast for relevant knowledge, if the risks are to be understood well enough to be ranked responsibly. For example, it is not enough to know how toxic a chemical is without knowing how many people are exposed to it. It is not enough to know the number exposed without knowing something about their identity (such as their ages, their degree of consent to the exposure, the benefits that they derive from the activities that impose the risk).

If the ultimate goal of ranking risks is to influence how they are managed, then there needs to be some notion of the control opportunities and costs. Even if control strategies will only be addressed systematically at some subsequent stage, there should be some chance for clear-cut "best buys" and "worst buys" to emerge. Particularly at the federal level, the political momentum needed to reallocate resources is more likely to be generated has revealed obvious misallocations of attention and resources.

Because the individuals performing the ranking cannot be expected to absorb more than a fraction of the relevant science, they will need succinct summaries of its pertinent features. Those summaries will have to consider each kind of risk (such as mortality, morbidity, birth defects) that is significant for any hazard in the comparison set. They will have to include both a best guess at the magnitude of each risk and an indication of the uncertainty surrounding it.

Finally, the numbers not only need to be correct, but also need to be understood as intended. As noted, there are well-documented problems with communicating technical information about risks. To some extent, these should be reduced within federal agencies with continuing interactions among technical staff having similar professional backgrounds. In such settings, conversational norms (which determine what is meant by potentially ambiguous words) should have had an opportunity to evolve. However, an essential goal of risk ranking is to force comparisons across such communities, not to mention outside agency boundaries. As a result, the opportunities for misunderstanding increase. Even within a technical group, adopting the common format needed to communicate with outsiders might be a source of confusion.

Ensuring the comprehensibility of risk estimates is an essential part of risk ranking. That requires attending to users' needs when designing information displays and empirically validating the effectiveness of the communication process. The structural similarity of the tasks faced by different federal agencies means that a set of general methods could be developed and evaluated, then adapted to specific circumstances. Although these tasks may require some additional effort, summarizing and explaining the scientific evidence regarding risks is part of the implicit or explicit mission of most agencies. If all that risk ranking does is help agencies to think systematically about the risks they manage and to describe these risks in a form that allows easy comparisons, then this method will have achieved something of considerable value.

Reflecting and Incorporating Public Values

The conclusions that emerge from a risk-ranking exercise should be credible reflections of public values. Achieving that goal at the federal level faces some substantial obstacles. It is hard to know what the public wants without asking, yet also very hard to ask. Ranking efforts based solely on expert judgment reflect a belief that our representative democracy is ill-suited to deriving from the general public a coherent set of preferences regarding the management of diverse risks. As a result, the explicit mandates imposed on agencies often seem ill-advised, while their general mandates do not provide the guidance needed to make hard choices.

An instinctive response to a need to know what the public thinks is to conduct a public opinion poll. Unfortunately, surveys are inappropriate for the kinds of issues faced by federal agencies hoping to prioritize risks. Conventional surveys cannot provide people with the background needed to understand complicated issues. They cannot provide the opportunity to reflect on issues and articulate stable beliefs. They have difficulty allowing people to express complicated and qualified beliefs. They do little to generate the public support needed to carry the day for proposals.

Surveys assume, in effect, that people have ready answers for whatever questions are posed to them. Common sense suggests that this is unlikely to be true; how could people instantly produce informed answers about everything? An extensive research literature shows the volatility of survey results possible from seemingly minor changes in question wording (Fischhoff 1991; NRC 1989; Turner and Martin 1984).

Surveys also assume that one can substitute an analytical representation of the public for actual people (Dryzek 1990). It is hard to point to cases where this strategy has proven effective once it is subjected to political scrutiny. Where issues matter, people want to be heard and not just counted. On some level, they realize the limitations of surveys and the imprecision of survey questions. They may resent having complex views squeezed through such an imperfect funnel.

The success of the recent risk-ranking exercises may be due, in part, to their having violated the assumptions that survey results can be substituted for real people. These efforts have allowed people to grapple directly with the issues over a period of time, and let them express their beliefs with whatever degree of precision is possible. Nonparticipants' perceptions of these processes do not seem to have been subject to empirical study. The hope is that the general public will be relatively satisfied with having people representing their interests taken seriously. Or, at least, the public will be more satisfied that way than with speaking directly, but imprecisely, through a survey (in which they have a small, but equal probability of being included).

At the federal level, these concerns require creating a forum for risk ranking where credible people are allowed to understand the issues, develop thoughtful opinions, and then stand behind what they say. Whether consensual opinions from such a process will emerge is an empirical question. If they do not, then there is little chance to generate the momentum needed to revise current risk management priorities. The recent emergence of varied risk prioritization efforts suggests that the time is ripe for some reform here (as with health care). It is an article of faith that reasonable people, treated respectfully, will agree about at least a few cases of risks that are receiving way too much attention and a few that are receiving way too little attention. The successes of the state and local ranking exercises gives some reason for optimism, allowing for the differences between trying to move a region and trying to move a nation.

Accommodating Current Institutional Arrangements

At the federal level, the responsibility for analysis and mangement of almost all risks falls within the institutional jurisdiction of one, and sometimes several, existing agencies. Those agencies constitute a reservoir—perhaps the only reservoir—of technical expertise for dealing with many risks. If priorities are to be changed, then these agencies will need to supervise the implementation of the changes. Agencies will have to rewrite their regulations, suggest changes in their enabling legislation, and cope with conflicting court rulings. They are likely to host and provide technical support for ranking exercises. They are likely to be required to perform rankings for internal management purposes. Doing that homework is a likely precursor to public involvement, insofar as the agencies need to get their own houses in order before inviting outsiders to reflect on the data that they have assembled.

On the other hand, the agencies cannot resolve everything by themselves. As mentioned, they do not have a clear signal from the public regarding how to address these issues. Furthermore, they are not indifferent to what conclusions are reached. To a first approximation, their budgets reflect the importance afforded to their risks. Changes in risk priorities would mean changes in budgets, both within and across agencies. As a result, there should be some natural inertia. It can express itself both directly, in the form of lobbying for current priorities, and indirectly, in the form of having better evidence regarding currently "favored" risks.

As a result, a risk-ranking strategy will have to provide a role for the agencies that exploits their expertise, balances their concerns with those of the public that they hope to serve, and helps them to make the transition to implementing the new priorities. It will have to allow them to provide enough structure to the problem for deliberations to be meaningful. However, the agencies cannot provide so much structure as to predetermine its conclusions.

DESIGN STRATEGY

The analyses presented in the preceding sections have led to a design strategy for the proposed risk-ranking procedure, a strategy that is described briefly here and elaborated further in the next section. This section considers both the analytical framework and processes of risk ranking and the social and institutional framework and processes needed to deal with the analytical issues. The analytical framework includes choosing risk categories and attributes, characterizing risks in terms of these attributes, and then performing the ranking and summarizing the results. The social and institutional framework includes an interagency risk-ranking task force, other risk-ranking groups, and the procedures they should use to perform their tasks.

An Analytical Framework for Risk Ranking

The risk-ranking process should begin at the level of individual agencies or of programs within agencies. Doing so will help achieve several goals:

- getting agency personnel to buy into the process, as something that they have had a hand in creating, rather than as something that was imposed upon them;
- getting the data about specific risks into workable form, relying on those who know them best;
- focusing on sets of risks that have relatively comparable effects; and
- increasing the chances of action by reducing jurisdictional conflicts.

Subsequent efforts could rank risks involving different programs within individual agencies as well as across agencies. The work done at each level will be designed to be compatible with that at other levels.

Even at the lowest level, a federal agency may face a very large number of different risks. Clearly, ranking that many risks is an impossible task. As a result, the first analytical challenge is bundling risks into roughly comparable categories. The detailed discussion of the proposed procedure itself provides a general framework that agencies could apply to create categories suited to the risks in their domain. In order to ensure that reasonable attention can be paid to each category, each agency should create no more than about thirty risk categories, since dealing with much larger numbers will pose serious procedural and cognitive problems for the rankers.

Each category should be relatively homogeneous in terms of the risk attributes of its members. These attributes are the features of hazards that may make people care about their risks. They might include the expected number of deaths per year, the expected number of injuries per year, the fraction of the population exposed to the risk, the degree to which exposure is voluntary, the expected amount of ecological damage, the reversibility of that damage, and so on.

Unlike the risk categories, which should be unique to agencies and programs, risk attributes should be the same in all cases. That is essential to allowing cross-agency comparisons. The second part of the procedure proposes a canonical set of risk attributes. It builds on risk perception research showing that people respond similarly to attributes within each of several clusters (Fischhoff and others 1978; Slovic 1987; Slovic and others 1979, 1980). We propose that one or two marker attributes be selected from each of these clusters.

Once the risk categories and attributes have been selected, the risks in each category should be characterized on each attribute. This risk characterization process will summarize the scientific evidence about each risk, including the degree of uncertainty surrounding it. The third section of the procedure discusses this process and presents one possible summary reporting form. In many cases, agencies will have the requisite information on hand, and will only need to translate it into the standard format. In other cases, additional analysis will be needed.

Once hazards have been organized into categories and characterized on a common set of attributes, the risk-ranking process can begin. This is the most explicitly subjective step in the process. Participants are free to assign whatever weight they wish to each attribute, including no weight at all. They are also free to treat the scientific uncertainty surrounding different estimates in whatever way they wish. These freedoms make possible a technical workup of risk data to guide the ranking process, without dictating its results.

Risk ranking is such a novel and complex task that the rankers will need help, if they are to develop and articulate accurate expressions of their beliefs. The fifth section of the procedure proposes a ranking elicitation process. It is designed to accommodate both the nature of the scientific data involved and the psychological processes required to evaluate them. This process uses multiple methods, in order to help rankers understand their task thoroughly and triangulate on their answers. The process is also iterative, in order to allow rankers to reflect on and refine their answers.

The final section of the proposed procedure describes procedures for summarizing the results of the ranking exercises. These procedures are intended to allow comparison of the rankings that different people and groups produce for the same risks, as well as the integration of rankings for different risks. They will reflect the degree of consensus within and across groups, as well as the degree of confidence felt by individual members. These measures of definitiveness are essential to understanding how committed rankers are to their conclusions, how much discretion is left to those who would act on the rankings, and what opportunities exist for increasing convergence (for instance, by creating missing data). Particularly at the federal level, risk ranking is likely to be an ongoing process. Such rich summaries are designed to facilitate that process, rather than to pretend to have produced results with an unrealistic degree of definitiveness.

An Institutional Framework for Risk Ranking

The institutional focus of a risk-ranking effort based on this proposal should be an *interagency risk-ranking task force* created early in the process. Such a group will draw members from all candidate agencies, not just those that will participate during the first round of ranking exercises, and will have responsibility for coordinating the overall risk-ranking effort. (Many of the tasks of such a task force are described briefly in the following paragraphs and thereafter in more detail in the section on the proposed procedure.)

The responsibilities of the task force will include ensuring the compatibility of the outputs of risk-ranking efforts (both within and across agencies), supervising the collection and dissemination of the lessons learned in different places, and commissioning studies that will serve multiple ranking efforts. The task force will have to direct agency staff members in ways that exploit (and respect) their expertise, without allowing them to dominate the proceedings. It will have to orchestrate the recruitment and continuing participation of the nonagency people needed to give the process credibility.

The creation of risk categories (elaborated in the section of the procedure on defining and categorizing risks) will be conducted primarily by agency staff, using guidelines promulgated by the task force that will review its proposals. These staff proposals will be subject to override by the risk rankers. However, the technical nature of the work justifies having a first approximation of the proposed categories prepared in advance. Because, unlike the attributes, the categories are specific to each agency, revisions should not complicate subsequent cros-agency comparisons.

The task force, in consultation with agency staff, will determine the canonical classes of risk attributes (elaborated in the section on identifying risk attributes), as well as propose a few "marker variables" that can act as surrogate descriptions for the larger number of variables in each class. These proposals also will be subject to revision by the risk rankers. If these rankers disagree about the choice of marker attributes, then that disagreement might be treated as a matter of taste that should not dramatically affect the results of their work. If rankers decide that an attribute class is irrelevant, then that decision is equivalent to assigning it a weight of zero, within the existing framework. Unlike the case of categories, if rankers decide to add or modify an attribute, that decision could disrupt comparability across rankings, meaning that the task force would have to accommodate its implications.

Characterizing the categories of risks in terms of the classes of attributes (elaborated in the section of the procedure on describing the risks with the attributes) is a matter for staff work. That characterization may require some negotiations to develop a version of the standard reporting format that staff are comfortable using. For example, in some cases, they may be unaccustomed to providing quantitative expressions of uncertainty and may need assistance in learning how this can be done. In other cases, the staff may be reluctant to describe the overall quality of the science (and the disagreements among different experts) and may need help in producing frank assessments that will not embroil them in controversies.

A critical process in this institutional work is choosing the people who will perform the rankings. Specifically, we propose that four different groups address each ranking task: a group of federal agency risk managers, two groups of laypeople, and a group of state and local risk managers. We also propose that the federal group include a small number of individuals from other agencies, in order to promote learning and convergence across the ranking exercises. The lay panels are conceptualized as being akin to juries, socially heterogeneous groups of citizens who are chosen for their independent opinions (rather than, say, as representatives of special interest groups). In order to allow diverse participation, lay members will be paid for their investment of preparation and meeting time.

In order to achieve the maximum degree of ownership over its conclusions, each group should manage its own affairs to the greatest degree possible. As a result, the role of agency staff is envisioned as providing an initial overview of the procedure as well as continuing technical support. For example, they can identify factual errors needing correction, point to value disagreements needing clarification, and retrieve needed scientific results. In order to increase the chances of success, experimental tests (and refinements) of the procedure should precede any applications. If outside groups (such as the League of Women Voters, the Society for Risk Analysis) wish to conduct parallel efforts using this framework, assistance should be extended by the agency staff to such a group.

Staff will draft written summaries of group results, circulate them for critical review between meetings, and present them to the interagency task force (along with group members, should that be desired). Final reports will reflect the range of opinion, as well as incorporate individual statements (as desired). Reports will be submitted to the interagency task force. The task force will maintain continuing contact with each ranking group to guarantee faithfulness to each group's conclusions and supporting arguments.

A RISK-RANKING PROCEDURE FOR FEDERAL AGENCIES

This section provides a detailed step-by-step discussion of the risk-ranking procedure, organized as follows:
1. Categorizing the risks that will be ranked
2. Identifying the risk attributes that should be considered
3. Describing the risks in terms of the attributes
4. Selecting the groups that will do the ranking
5. Performing the rankings and merging the results of the rankings
6. Summarizing the results in detail for use by the general public and risk decisionmakers

The first two steps will be performed simultaneously. Before they can be ranked, risks must be sorted into categories (radon in homes, lead in homes, fire in homes, and so forth) and the set of attributes must be chosen that will be considered in evaluating them (number of deaths/year, equity of exposure, controllability of exposure, and so forth). We begin by discussing the problem of sorting risks into categories.

Categorizing the Risks That Will Be Ranked

In order to rank risks in a meaningful way, they must first be sorted into a modest number of roughly comparable *risk categories*. These are the broad categories (no more than about thirty in number) into which the risks that an agency manages are sorted for purposes of ranking. It is these categories that will be ranked.

Given the diversity of the risks faced by different federal risk management agencies, no single categorization scheme is likely to serve all their needs. As a result, each agency should devise its own set according to the following criteria to ensure their logical soundness. Each set of categories should be a simple, well-defined partition of the set of risks, easily understood by nonexpert users and balanced in terms of the relative size of the categories.

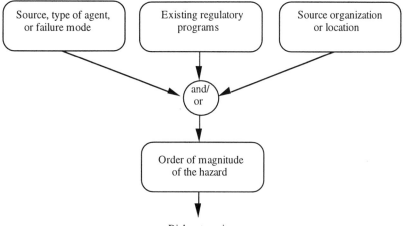

Figure 1. Constructing Risk Categories: An Alternative Approach.

In *Unfinished Business*, EPA sorted environmental risks primarily according to the physical (heat, sound) or chemical (carbon dioxide, mercury, radon) processes that actually cause the harm to humans or the environment. While this categorization scheme works well for many environmental risks, it is not a sufficient rule for categorizing the risks managed by some agencies.

We recommend that agencies consider the scheme summarized in Figure 1 for constructing their risk categories.

The first instinct of many agencies will be to sort risks according to *existing regulatory programs*. This categorization has the clear advantage of mapping directly on to administrative units within an agency. It is often highly correlated with the "source, type of agent, or failure mode" (explained below) that causes the risk. For example, at EPA, the Office of Air Quality Planning and Standards is responsible for the various "criteria air pollutants."

As mentioned, the physical or chemical process actually causing harm is another obvious categorization principle. We conceptualize this in terms of source, type of agent, or failure mode. *Source* refers to the activity that imposes an environmental loading or other event that may lead to the imposition of a risk. *Agent* refers to the kind of loading. *Failure mode* refers to events that cause risks to be realized. For example, at EPA, air pollution could be sorted in terms of the source of emissions (power plants, motor vehicles, and so forth) or by the type of pollutant (such as sulfur dioxide, nitrogen oxides, ozone, particulates). For the use of the Federal Aviation Administration (FAA), air crashes could be sorted by accidents on the ground, accidents in terminal areas, and accidents en route. Terminal area accidents could be further sorted by such failure modes as wind shear, icing, and so forth.

Sometimes, it may make sense to sort risks in terms of source organization or location. This is most likely when only one or a few organizations or locations are responsible for the hazard. For example, the Department of Energy (DOE) might want to differentiate the Hanford site from the Oak Ridge site. In contrast, the FAA probably does not want to regulate risk in, say, the operations of USAir differently from the operations of United or American Airlines.

The *order of magnitude* of the hazard may seem like a circular criterion. If the point of the exercise is to rank risks, why would one want to categorize them in terms of how big they are? The point of explicitly considering magnitude is to avoid ending up with categories that contain too many large risks or that lump very high and extremely low risks. If categories violate these conditions, they should be reorganized in order to differentiate among the hazards in large categories and to find common bases for hazards in small categories.

This framework for categorizing risks should be viewed as a checklist of things to consider when developing categories. Some agencies may find it sufficient to operate with a single criterion. Others will want to incorporate two or three criteria. For example, suppose that DOE were to create a risk category called "plutonium at DOE sites." The risks associated with plutonium stored in different kinds of structure (fortified bunkers, cinder block buildings in tornado country, shallow underground tanks of varying integrity) are significantly different. In this case, it probably makes sense to use at least two other factors (location/structure type and expected order of magnitude of the hazard).

An iterative approach is essential. Once an agency has defined a tentative set of categories, it should subject them to critical review. Once a final set has been chosen, agencies should work hard to stick with them since ad hoc reclassifications may result in double counting of risks, placing risks in overlapping categories, or missing important categories.

Here are two examples of how agencies might categorize the risks for which they are responsible:

- The FAA might decide to categorize by source organization and by failure mode (type of agent). Its sources might fall into civilian and commercial aircraft. Examples of failure modes might include icing problems, wind shear, engine fires, loss of communication, and so forth. If so, one category of risks to be ranked by FAA might be: wind shear accidents involving commercial aircraft.

- The Consumer Product Safety Commission (CPSC) might decide to categorize risks both by source (that is, by product) and by type of agent/failure mode. Source categories could include clothing, toys, electrical appliances, microwaves, computers, detergents and cleaning products, and so forth. Types of agents/failure modes could include ingestion of poison/toxin, mechanical trauma, ionizing radiation, fire/burn/shock, and so forth. If so, one category of risks to be ranked by CPSC might be: defective wiring in electrical appliances (such as irons or power drills) causing fires/burns/shocks.

Because of the extreme difficulty of ranking large numbers of separate risks, it is important to try to produce a list that can fit comfortably on a single piece of paper (say, twenty to forty items). When this poses difficulties, one option is to build a hierarchical or "nested" classification scheme. For example, "risks from Superfund sites" might be broken into a variety of more-detailed risks by agent, location, and expected order of magnitude of the hazard. Even if these subcategories are not fully ranked by all the groups, there will at least be explicit recognition that, while the overall category may rank at one level, specific subparts may deserve more or less attention or concern than this overall rank implies.

Identify the Risk Attributes That Should Be Considered

Risk is a complex concept. Expected numbers of deaths and injuries are clearly important, but a variety of other considerations, or *risk attributes*, such as degree of controllability, may also matter. To the extent possible, these attributes should be:
- comprehensive (to ensure that nothing important has been left out);
- nonredundant (to avoid double-counting);
- preferentially independent[2] (to allow simpler aggregation methods);
- measurable (to allow explicit and consistent estimates); and
- minimal in number (to reduce complexity).

Because some of these criteria (such as comprehensive, minimal in number) can conflict, the process of choosing attributes will have to involve judicious compromises, the guidelines for which are described below.

Because different agencies are responsible for managing different types of risks, they are likely to have different views about how these trade-offs should be made. At the same time, it is important that the risk attributes used by each agency be the same, in order to make interagency comparisons possible.

The empirical literature on risk perception provides ways to resolve this dilemma (for example, Fischhoff, Watson, and Hope 1984; Slovic and others 1979, 1980). In it, researchers have asked both laypeople and technical experts to evaluate long lists of risks in terms of a wide variety of attributes. Using factor analysis, these studies have found that attributes can be roughly sorted into three sets (or factors), which we will label "number," "knowledge," and "dread."[3] For any given hazard, people's judgments about attributes that fall *within* one of these sets tend to be high or low together, that is, they show high interattribute correlations. In contrast, comparisons of attributes that lie in different sets do not tend to go up and down together, that is, they display low interattribute correlations. Thus, as long as the attributes used in the ranking process include a few from each set, the results of the ranking process are not likely to depend very much on which specific attributes are used. In the paragraphs that follow, we discuss how attributes from each set can be operationalized, so that agencies can assign numbers to them for use in their ranking exercise. Because environmental considerations are important in many risks, we have elaborated the original groupings developed by Slovic and others (1980) of "number," "knowledge," and "dread" into four sets of attributes:

- Attribute Set 1: Numbers of deaths and injuries
- Attribute Set 2: Extent of environmental impact
- Attribute Set 3: Knowledge
- Attribute Set 4: Dread

Rather than select the actual attributes to be used, we propose that the interagency risk-ranking task force select from each of these sets one or two attributes that all agencies agree will meet their needs. These selected attributes will serve as representative markers for all the attributes in the set. In order for this selection process to be well informed, it should be done after at least several agencies have already developed draft lists of their risk categories and circulated them to the interagency risk-ranking task force.

For each attribute in the list, we have specified the units to be used as well as provided some indication of the precision with which we believe it will be necessary to estimate that measure. In some cases, the necessary precision may depend on the specifics of the risk categories.

In several cases, we have suggested constructed scales using qualitative measures. In these cases, we have usually given a very simple illustration of how we expect the scale to be applied. Working out more detailed rules for the use of constructed scales should be the responsibility of the interagency risk-ranking task force insofar as they represent the groups that will have to actually use them.

Attribute Set 1: Numbers of Deaths and Injuries. The numbers of people killed or injured by a risk can be characterized in a number of different ways. Different measures carry different implications. Attributes in this set include:

- *Expected number of annual fatalities* (units: deaths/year; resolution: a factor of 10). This measure is probably the most easily operationalized, since for many risks reasonably good estimates are already available. Using expected number of deaths as a metric makes two important assumptions: that every life (regardless of age, health status, productivity, and so forth) should count the same and that the expected value, rather than some more conservative metric (such as the upper 90th percentile of the probability distribution in annual risk) is the appropriate measure. The decision analysis literature provides arguments for using expected values. Some other features of mortality risk are incorporated in other attributes.

- *Expected number of annual nonfatal casualties, including illness and injuries* (units: cases/year; resolution: a factor of 10). This measure allows one to account for human health effects that are serious, but not fatal. In order to operationalize this attribute, one must develop a threshold for what qualifies as an "injury." The same qualifications listed under expected number of annual fatalities also apply here.

- *Expected number of annual person-years lost due to death* (units: person-years; resolution: a factor of 100). While more difficult to compute, this attribute has the advantage (relative to expected number of annual fatalities) of reflecting age of death. Thus deaths in old age count far less than deaths in childhood.

- *Expected number of annual person-years lost due to nonfatal casualties, including illness and injury* (units: person-years; resolution: a factor of 100). While more difficult to compute, this attribute has the advantage (relative to expected number of annual injuries) of reflecting the duration of the debilitating effects. Thus, injuries that permanently disable will count more than those from which victims can soon recover.

- *Expected number of annual person-years lost* (units: person-years; resolution: a factor of 100). This measure is operationalized as the sum of person-years lost due to deaths and injuries. Its advantage is capturing in a single measure time lost due to both deaths and injuries.

In each of these measures, it is possible to weight differentially the individuals whose life and health are at risk (in addition to the differential weighting of young and old reflected in "person-years lost"). For example, concerns of environmental equity might prompt giving greater weight to risks borne by the poor.

Attribute Set 2: Extent of Environmental Impact. Risks to the ecosystems and the natural environment are central to some agencies' tasks and the concerns of many citizens. Some aspects of these risks fit under the knowledge and dread attribute sets, but neither adequately captures the extent of the risk.[4] Several measures are proposed:

- *Area affected by ecosystem stress or change* (units: square kilometers; resolution: a factor of 10). Application of this measure requires the establishment of a threshold, below which effects will be considered negligible. As part of this threshold, it may also be desirable to exclude regions under certain types of land use, such as those that are already heavily industrialized.

- *Magnitude of environmental impact* (units: constructed scale—negligible, modest, large). Examples of levels of impact should be developed to provide guidance on how to evaluate impacts in terms of a three-level constructed scale: negligible, modest, and large. If the area measure is not also adopted, then this measure may need to be area weighted (that is, for a given level of seriousness, assign more weight to impacts that have larger spatial extent so that a "large" ecological impact that covers the entire state of Montana gets ranked above a "large" ecological impact that involves a single half-acre plot).

- *Special regions/resources impacted* (units: constructed scale—negligible, modest, large). A list of special regions of concern, such as national parks, wildlife refuges, and so forth, should be constructed. Examples of levels of impact should be developed to provide guidance on how to evaluate impacts in terms of a three-level constructed scale: negligible, modest, and large. If the area measure is not also adopted, then this measure may need to be area weighted.

- *Expected numbers of species lost* (units: species; resolution: a factor of 10).

Attribute Set 3: Knowledge. A number of attributes studied by Slovic and others (1979, 1980) involve various aspects of lay and expert knowledge of risks and their properties. Attributes in this set include:

- Degree to which risk is *observable* (units: constructed scale—No, With difficulty, With ease). A measure of how easily those at risk can observe that they are at risk.

- Degree to which risk is *known* (units: constructed scale—No, To some (less than 50% of those at risk), To most). A measure of how well known the risk is within the group of people who are at risk.

- Degree to which impacts are *delayed* (units: constructed scale—<1 year, 1–10 years, >10 years). A measure of how much time typically goes by between the time that exposure occurs and when risk consequences result. If this attribute is not selected, then one may want to modify the expected number of death/injury attributes to incorporate some time discounting in case one wants to weigh delayed impacts differently from immediate impacts.

- Degree to which risk is *reversible* (units: constructed scale—Yes, Frequently, No). A measure of how permanent the impact is once it has occurred. This measure is especially useful for considerations of ecological risk and for nonfatal casualties.
- Degree to which risk is *new* (units: constructed scale—<10 years, 10–50 years, >50 years). A measure of when the technology or activity that constitutes the hazard was introduced.
- Degree of *scientific understanding* (units: constructed scale—High, Medium, Low). A measure of how well the scientific community understands the risk. For example, auto accidents should be classified as high, conventional air pollution as medium, and 60-Hertz fields as low.

Attribute Set 4: Dread. Dread is a general term Slovic and others (1979, 1980) have used to characterize a variety of factors which include:
- *Individual controllability* (units: constructed scale—Low, Medium, High). A measure of the degree to which an individual can control his/her exposure to the hazard and/or its subsequent effects. For example, food poisoning from commercially packaged foods should be classified as low, auto accidents should be classified as medium, and motorcycle accidents should be classified as high. An alternative measure called *degree of voluntariness* might be introduced. While voluntariness captures somewhat different ethical principles, the two measures are strongly correlated.
- *Population extent* (units: percent; resolution: nearest 10%). A measure defined as the proportion of U.S. population potentially at risk.
- *Catastrophic potential* (units: dimensionless ratio; resolution: a factor of 100). A measure defined as the ratio of the upper 99% worst case impact to the annual expected impact. Some risks hold potential for large disasters. This measure is designed to operationalize that potential. "Impact" can be any of the measures of mortality/morbidity and/or of environmental impact listed in Attribute Set 1 or 2.
- *Outcome equity* (units: constructed scale; resolution: a factor of 3). A measure defined as the number of those who receive benefits from the activity divided by half the sum of the number who receive benefits and the number who are at risk. It is probably sufficient to sort results into three categories (High = 1–3, Medium = 4–10, Low >10).
- *Intergenerational risk* (units: constructed scale—Negligible, Modest, Large). A measure of the extent to which the risk has direct impacts

on future generations. For example, it might include a genetic effect that will only be realized when exposed parties have children.

- *Controllability* (units: constructed scale—Low, Medium, High). A measure of how well the risk can be controlled through reasonable and affordable social actions. "Low" implies that reductions of greater than 5% in risk level cannot be achieved with expenditures of $105 per death per year. "Medium" implies that reductions of 5% to 30% in risk level can be achieved with expenditures of $105 per death per year. "High" implies that reductions of greater than 30% in risk level can be achieved with expenditures of $105 per death per year.

Describing the Risks in Terms of the Attributes: Summary Forms

Once a stable set of categories has been created by each risk-ranking group, and a set of attributes has been accepted and defined, staff at each agency should produce a *risk summary form* for each risk category. The form should define concisely each risk category and provide the following:

- A nontechnical, qualitative description of the risk. This should include a brief category name and a short paragraph explaining what the risk category includes.
- A quantitative evaluation of the risk in terms of each chosen attribute. These quanitative values should take the form of "best estimates." When the range of uncertainty associated with this best estimate is wider than the specified resolution, the associated uncertainty should be reported in the form of a 95% confidence interval.[5]
- A brief qualitative discussion of the state of scientific understanding and any special circumstances.

Because they will be used by people who are not experts, these risk summary forms should be written in language that is no more technical than that used in *Popular Science* or *Popular Mechanics*.

The forms should be small enough to be easily handled and sorted, but large enough to allow readable text. In order to illustrate the tabular summary, we must assume a particular set of risk attributes. For a given risk, the tabular summary thus might look like that illustrated in Figure 2.

Selecting the Groups to Do the Ranking

The objective here is to select groups of people, capable of ranking the risks managed by a federal agency. The groups—risk experts, nonexperts, risk managers—will be given the agency's risk categories, a char-

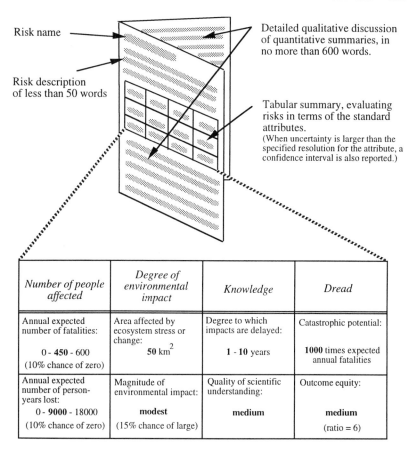

Risk name

Detailed qualitative discussion of quantitative summaries, in no more than 600 words.

Risk description of less than 50 words

Tabular summary, evaluating risks in terms of the standard attributes.
(When uncertainty is larger than the specified resolution for the attribute, a confidence interval is also reported.)

Number of people affected	Degree of environmental impact	Knowledge	Dread
Annual expected number of fatalities: 0 - **450** - 600 (10% chance of zero)	Area affected by ecosystem stress or change: **50 km**^{2}	Degree to which impacts are delayed: **1 - 10** years	Catastrophic potential: **1000** times expected annual fatalities
Annual expected number of person-years lost: 0 - **9000** - 18000 (10% chance of zero)	Magnitude of environmental impact: **modest** (15% chance of large)	Quality of scientific understanding: **medium**	Outcome equity: **medium** (ratio = 6)

Figure 2. Risk Summary Sheet: Suggested Format

acterization of the risks in terms of the focal attributes, and assistance with the methodology of ranking.

Risk ranking requires value judgments. As a result, it matters who does the ranking, unlike scientific analyses that should come out the same regardless of who performs them. We believe that more than one group is needed in order to increase public confidence that the results are robust and not just the product of particular group dynamics. At the same time, having too many groups may complicate merging their conclusions into a summary ranking. The dynamics of individual groups are likely to be best if members enter with similar levels of expertise. All should be afforded equal respect.

In our proposed system, both expert and lay groups will perform rankings. To the extent that the two types of groups produce similar rankings, that would add confidence to the conclusions. To the extent that experts and laypeople disagree, significant differences have been revealed between their beliefs and/or values. Such differences need to be carefully explored and understood.

Sometimes, the differences may result from public misunderstanding; if so, then more effective risk communication might help. In other cases, the differences may result from risk managers, giving insufficient weight to values that are important to the public; if so, then their values need to be incorporated in management. In still other cases, lay group members may have defensible skepticism regarding the confidence that technical experts place in risk analyses.

In addition to risk experts and members of the general public, a third group has unique insights relevant to risk ranking. These are officials on the firing line who are charged with implementing federal risk management decisions at the state and local levels. As people involved in the day-to-day tasks of managing risks, these officials bring a unique perspective that marries public values and concerns with a practical understanding of policy implementation. Moreover, practices mandated at the federal level may constrain their ability to establish reasonable priorities at their levels.

Grouping the Risk Rankers. With these considerations in mind, we propose convening four groups to perform rankings:

- One group of *federal agency risk experts and managers*. Most members should be drawn from the staff of the agency doing the ranking, with a small number from the staffs of other federal agencies in order to promote cross-fertilization among federal risk management agencies. In some cases, it may be desirable to supplement the group with a few outside experts. Composition of this group should be determined by agency management.

- Two groups of *laypeople* selected to reflect the diversity of the United States. In order to understand and perform the ranking tasks effectively, they will likely need to have at least a solid high school education. In order to have a free discussion, members of this group should *not* be representatives of formal advocacy groups. As discussed below, we propose that these two groups be selected by a contractor through a quasi-random process.

- One group of *state and local risk managers*. This group should consist of people like state risk management officials, local enforcement officials, and other appointed and elected officials. Federal agency risk management staffs with frequent contacts should be a good source of nominations. As discussed below, actual selection of the group should be the responsibility of a contractor.

Each of these four groups should have between ten and fifteen members in order to balance the needs for diversity and manageability. While the groups may need some staff support (for such work as analytical help), their credibility will be greater to the extent that they manage their own affairs. As a result, organizers should appoint two group members, who have had experience running meetings, to serve as chair and vice-chair. They should receive careful instructions (and perhaps some preparatory training) in what their group is expected to do.[6] One staff member should be identified as someone who can help the group out if it encounters problems, as well as to be an authoritative source on the entire process.

Choosing and Organizing the Risk Rankers. In principle, one might choose the panels by drawing a random national sample and screening candidates for their ability to handle the material. In practice, though, that seems like a clumsy and inefficient way to secure the sort of credible representative panel that is needed.

Instead, we propose that a contractor manage the recruitment process. Because most lay participants will have jobs, meetings will have to be held on evenings and weekends. As a result, the groups should be organized on a local basis. The contractor should devise a procedure to select two locales, separated by at least 500 miles, which are culturally different. (For agencies that divide the country into different administrative regions, the two locales should fall in different regions.) For each locale, the contractor should construct a statistical profile of the age, racial, educational, and economic makeup of the community. A statistically representative fifteen-person group should then be constructed. Community groups, such as nonpolitical clubs, churches, and parent-teacher associations that serve the relevant communities, should be approached to obtain suggestions for possible candidates. Candidates should be briefly interviewed to establish their willingness to participate, to confirm basic data such as educational level, and to determine that they are not active members of any advocacy group with positions on the risks that will be considered. At least two male and two female candidates should be identified for each of the fifteen openings. Final membership would then be determined by random draws for each position. Starting with fifteen members should assure a manageable group size, while leaving room for some inevitable attrition.

Because a significant amount of preparation and meeting time will be required over a period of some months, members should be paid for their participation (perhaps $100/day). In addition, when transportation or child or elder care presents problems, relevant support should be provided.

A contractor should also organize and run the group comprised of state and local risk managers. Again, economic and logistical reasons will probably require confining this group to a region, although a few members might be able to come from long distances. The agency should provide fifteen nominations, while the contractor independently obtains an additional fifteen nominations by consulting experts and practitioners. All candidates should be current office holders or should have held an appointed or elected state or local risk management office within the past five years. Candidates should be contacted to determine their willingness to participate if selected. Final selection of a group of fifteen should be made by a random drawing of names from the assembled candidate pool of thirty.

As discussed in the section on implementing this proposed procedure, the process should be managed in such a way to allow groups to conduct their ranking deliberations in private so as to allow the freest possible exchange of ideas. It would be inappropriate for interest groups to be tracking the proceedings and lobbying members for specific outcomes.

Performing the Rankings and Merging the Results

Behavioral researchers have found that the values that people express can be highly sensitive to seemingly irrelevant variations in how questions are posed—in situations where they have not formulated well-articulated answers in advance. Given the novelty of risk ranking, responses to it could show such inappropriate sensitivity (or "lability," as it is sometimes called). The best way of dealing with this threat to the validity of value measurement is to use multiple methods, compare the results, and then, if possible, resolve the inconsistencies among them. To this end, we propose that each group employ two quite different procedures. The group as a whole will rank the risks holistically. However, before that process begins in earnest, each individual member will privately complete an attribute-weighting task, based on procedures from multiattribute utility theory. The attribute-weighting task has two purposes. The first is to help focus individual panel members on the task and to sharpen their personal values, prior to beginning the group process. The second is to serve as a check on unhealthy group dynamics, by preserving a record of independently derived judgments, produced before the intervention of any domineering group members or collective myopia. Either at the group or the individual level, the best protection against allowing one perspective to have undue influence is to invoke multiple perspectives explicitly. If the attribute-weighting task and the holistic ranking produce different results, then participants can decide how to resolve the apparent contradictions. While this may take more time, to the extent that consensus can be achieved, the results will be much more robust. Additional details on these two procedures are provided at the end of this section.

Suggested Series of Meetings for Risk Ranking. We propose that the groups proceed with their work in a series of five meetings, as described below, with each receiving the same materials and instructions:

Meeting 1: General organization (less than a day)
- Explain the exercise generally.
- Allow members to get acquainted.
- Briefly describe the risk categories.
- Discuss the risk attributes that will be used at some length.
- Explain the process in greater detail.
- Distribute the four-page risk summary sheet for each risk to be ranked.

Allow the group several weeks to read and reflect on the risk summary sheets.

Meeting 2: Briefings and first ranking (two to three days)
- Provide briefings on each risk category and answer questions from group members.
- Have each group member individually complete the attribute-weighting task.
- Receive a briefing on the holistic ranking procedure.
- Begin to generate consensual holistic rankings of all the risks.

Meeting 3: Complete ranking (two to three days)
- Review and continue to refine consensual holistic rankings of all the risks.
- When the group feels that it is getting close to a holistic ranking, it should be presented with rankings based on an analysis of the values expressed in the attribute-weighting task (of Meeting 2). Participants should compare these rankings, using them as a prompt for reconsideration and further discussion.
- Revise the holistic ranking (or rankings, if the group is unable to reach consensus).
- With staff assistance, prepare a fairly rich description of the considerations that led to the results and report the results.

Once each group has completed these steps, the summary documents from the four groups will be distributed to each member of each group.

Meeting 4: Final within-group discussion (one day)

- Discuss the reports on the deliberations and conclusions reached by the other three groups.
- Revise the group's own document, if serious oversights are noted.
- Elect four representatives to attend the final intergroup synthesis meeting.
- Provide the representatives with detailed advice on what weight to assign different issues in a subsequent meeting devoted to reaching consensus among the four groups.

The final meeting will involve a sixteen-person panel comprised of four representatives from each of the four ranking groups. We believe that this meeting, too, would be best if run primarily as a closed meeting. Doing so will facilitate the free exchange of views. It should be particularly helpful in allowing lay members to speak their minds, despite the presence of experts and professional risk managers.

Meeting 5: Intergroup synthesis (three days)

- Using holistic ranking procedures, the meeting should work for a day to resolve differences in the four sets of rankings.
- When it seems to have identified irreconcilable differences, the panel should work to articulate the reasons for their disagreement.
- Overnight and during the next morning, staff (and possibly individual committee members) should prepare a draft report that the committee should review and edit during the afternoon of the second day and morning of the third day.
- At midday on the third day, the meeting should be opened to the public. The synthesis report should be presented and a representative from each of the four groups should have an opportunity to make comments. The synthesis report should include the four separate group reports as its first appendix.

After the final meeting, the synthesis report should be circulated to all participants in the four groups. Each should have an opportunity to write additional individual comments, to be included in a second appendix when the final report is published.

Implementing Two Ranking Procedures: Holistic vs. Attribute-Weighting. We provide here some preliminary suggestions on how the attribute-weighting and holistic ranking procedures might be implemented. However, at this stage these details should be viewed as simply illustrative.

While there are theoretical bases in the literatures on behavioral decision theory, decision analysis, and multiattribute utility theory on which to base the design of such procedures, more detailed work is needed in order to create viable methods. Those methods should be evaluated empirically before being implemented in such a critical enterprise. There are a number of groups in the country with the requisite background to perform these design and evaluation tasks. However, we stress that this is not a task that can be handed to the average beltway consulting firm.

One feature of the holistic ranking procedure is its emphasis on getting things roughly right on a large scale. Given the difficulty that individuals, not to mention diverse groups, will face in this ranking task, it is unrealistic to expect a crisp ordering to emerge. A more realistic outcome is sorting the risk categories into a few broad classes paying particular attention to identifying those that deserve the highest—and the lowest—ranks. The group should not expend energy on the meaningless task of achieving precise ordinal ranks for risk categories whose ranks are broadly overlapping.

Holistic Ranking Procedure. While the details should be refined through some experimental work, the holistic ranking procedure might operate in two steps. First, after carefully studying and being briefed on all the risks, each member should be asked to sort their risk summary forms into three piles (high, medium, and low). Then, the results should be summarized on a standard display that allows everyone to see how many people ranked each risk in each category.

Risks that are ranked similarly by all group members should be noted and set aside for the moment. Risks that involve considerable disagreements in ranking should be discussed to determine what is driving the differences, starting with those that involve the greatest disparities, such as cases in which the group is split between high and low.

After all the relevant risk categories have been discussed, the group should repoll members for each risk category and produce a new summary.

When the resulting new summary has been produced, those risks that were ranked in the high category should be listed on a ballot. Each group member should receive two or three times as many votes to distribute as there are risk categories on the ballot. Votes should be distributed in proportion to the relative importance they assign to each.[7] Results should be reported both for each individual and for the group as a whole.

The process should be repeated for the risk categories that have been ranked as medium and low. The reason for voting three times, as opposed to just once on all twenty or thirty risk categories, is that it is easier to make comparisons among smaller numbers of items and, at this stage, possible overlaps between risk categories in the three groups are not of great importance.

This is the point at which the first meeting should end.

At the beginning of the second meeting, the results of the attribute-weighting exercise (described below) should be presented for all the risk categories in a table that reports rank orders. These results should be compared with individual and group rank-order results obtained during the first session of holistic ranking. There are likely to be a number of significant differences. These differences should be identified and discussed in order to allow individuals and the group as a whole to reflect on how they believe the inconsistencies can most appropriately be reconciled.

The results of all previous rankings should be reviewed to identify groups of risk categories whose ranks are so similar (for example, because of uncertainties in estimated attribute values) that most participants find they are unable to differentiate ranks among them. These risk categories should be combined for ranking purposes so as to reduce the total number of items that must be considered in the final sort.

At this point, each participant should receive a set of cards that list the names of all the risk categories to be sorted. Risk categories that have been grouped together as having the same rank should all be listed on the same card. Working as a group, the committee should try first to sort the cards into three piles (high, medium, and low). Then they should try to order each pile, using pair-wise comparisons, beginning with the high-ranked cards. Participants who, after a serious try, find that they are simply unable to accept the results that are emerging from this group process should drop out and perform their own private rankings which can be reported as a minority opinion.

The Attribute-Weighting Procedure. In contrast to the holistic ranking procedure, in which risk categories are considered in their entirety and compared with one another, the attribute-weighting procedure focuses on the relative importance of each risk attribute. These weights are then applied to *calculate* the relative rank of each risk category.

Early in each group's deliberations, each participant will be asked to complete individually an attribute-weighting form. If the attributes can be considered to be independent, then the procedure is fairly straightforward. Participants can be asked to make comparisons between various attributes to help them begin thinking about them. Each participant could then be given a play-board that lists all the

attributes and 100 checkers to allocate across the attributes in proportion to how much weight they believe that each one should receive in
evaluating a risk. A few risks could then be ranked, as illustrations, to
see if the results "feel right" and then the participant could be given an
opportunity to reassess his or her weights if they wish.

If a participant's views about the attributes are not independent,
then things get somewhat more complicated (Keeney 1980; Keeney and
Raiffa 1976). Determining whether there is likely to be an independence
problem and, if so, just how much attention it should receive, is one of
the tasks that should be addressed by the contractor charged with
developing, empirically evaluating, and refining the ranking methods.

Once a set of weights (and utility function form) has been obtained
from each participant, staff can compute implied individual rankings
(together with appropriate sensitivity analyses) for each participant.
These are to be introduced during Meeting 3 to help participants develop
a second perspective on the ranks they have produced through the
holistic ranking procedure.

Summarizing the Results for Decisionmakers and the Public

The summary list of rank-ordered risk categories should prove useful
in educating government officials, corporate leaders, and members of
the general public, and, in time, influencing decisions regarding risk
management. That summary, however, is only part of the story that the
ranking exercises have to tell. The same list can mean different things if
it reflects either a strong consensus or a weak plurality of views. It can
motivate different actions if disagreements reflect conflicting values or
alternative beliefs regarding uncertain scientific evidence. As a result,
the summary report should discuss the value and fact issues on which
it was difficult or impossible to reach agreement, additional scientific
knowledge that would have made the task easier, and problematic procedural issues. These discussions should provide useful guidance to
federal risk managers and program officers as they plan scientific
research, policy research and development, and risk management.

In order to ensure candid descriptions of these issues, contractors
should be instructed to treat the entire process as a learning experience.
Too much emphasis on deriving a "final answer" will miss valuable
lessons. Contractors should prepare transcripts or take extensive computer-based notes as the sessions proceed and then edit the more valuable and enlightening portions of the material into the final reports. At
times, such final reports may include sanitized versions of actual participant quotations in order to illustrate particularly salient features of
the group discussions.

IMPLEMENTING THE RISK-RANKING PROCEDURE

If federal agencies undertake the risk-ranking procedure we have outlined, several coordinated steps will be required for its implementation:
1.   A research-oriented group that is expert in behavioral decision the-
     ory, decision analysis, and multiattribute utility theory should be
     contracted to develop, empirically evaluate, and refine the
     attribute-weighting and holistic ranking procedures that the rank-
     ing groups will use. This same set of procedures should then be
     used by all agencies. As part of their responsibilities, these investi-
     gators should train the necessary agency and contractor staff in the
     use of the procedures they develop.
2.   An interagency risk-ranking task force should be created to coordinate
     the overall procedure. This task force should include representation
     from as many federal risk management agencies as possible, not
     just those that will participate during the first round of ranking
     exercises.
3.   Each agency participating in the first round of risk ranking should
     form its own intra-agency risk-ranking group. That group should
     include, at a minimum, its representative to the interagency task
     force, a member of its technical staff, a member of its policy staff,
     and the chair of its internal ranking panel.
4.   One or more contracts should be let to organizations with the logis-
     tic capabilities to provide staff support and run the nonagency
     ranking groups. The use of contractors is suggested for reasons
     elaborated below.

Given the open meeting requirements imposed on government
agencies by the "Government in the Sunshine" law (5 U.S.C. §552b), we
recommend having the three nonagency ranking groups run by a contractor. Several firms are well qualified to carry out the actual staff work
of coordinating and conducting the rankings. Different agencies are
likely to want to use their own firms. Because of the applied nature of
the work and the tight time lines, a university group is probably not the
best choice for this phase of the work. University faculty could advise
on methods, review materials, and conduct independent evaluations of
the overall process. Each contractor should have substantial technical
skills and knowledge in the domain of the risks being considered. The
contractor should have a good working familiarity with the literatures
in risk assessment and decision theory, but should not be slavishly tied
to any paradigm. The contractor should have a demonstrated track
record of being able to work with and communicate to semi-technical
and nontechnical people on complex risk issues in a friendly and effec-

tive manner. The contractor should accept the principle of an independent evaluation of their procedures and products.

CONCLUSIONS: WHAT RISK RANKING CANNOT DO

If federal risk management agencies were to perform the kind of systematic ranking exercise that is proposed here, we believe that considerable direct and indirect benefits would result. However, it is important to recognize that ranking risks does not solve the risk management problem and that ranks may not translate directly into budgetary priorities. A ranking says nothing about what it will cost to do something about risks. Some low-ranking risks may be eliminated with very little cost or effort. Clearly, when this is so, they should not be ignored. At the same time, there may be high-ranking risks for which effective risk management is so expensive that only limited risk reduction is possible. In other words, federal agencies must consider the marginal cost of risk management when they allocate their own or others' resources to risk management. When risk rankings and risk actions diverge, agencies will have to explain why they have not neglected or rejected the conclusions of their ranking panels—lest ranking be perceived as an empty exercise.

Risk with very high and very low marginal costs of control are the exception, not the rule. Many risks can be significantly reduced at costs ranging from a few tens of thousands to a few million dollars per death averted. This is probably the same sort of range in expenditure levels that we can expect from shifts in political and institutional attention. Risk rankings of the kind that will result from the procedure we have proposed should prove a valuable tool for education and for guiding future risk decisions about managing such risks.

Finally, there are a number of important risks in our society that will *not* be considered in any first round of federal agency risk ranking. Partly, this will be because not all agencies will participate. Partly, this neglect will be because not all risks fit neatly into the domain of federal agencies. At some stage, the interagency risk-ranking task force would be well advised to discuss this fact, and explore mechanisms to fill this gap.

ENDNOTES

[1] There are important differences between comparative risk projects at the state and national levels. At the state level, it has been possible to use risk ranking as a tool for political consensus-building. For example, in Louisiana, a thirty-member Public Advisory Group was structured around advocacy groups, state agencies, academics, and medical professionals. Decisions are frequently

described as having been "negotiated." A prime goal was to use these opinion leaders to communicate the results of the committee's work back to their constituencies. At the federal level, interests are more diverse and it is unlikely that a small number of clearly definable constituencies can be identified. Thus, adding the requirement of interest group negotiation to the already complex risk-ranking task is unlikely to be productive at this level.

[2]*Preferentially independent* as used here means that a person's judgment about the value of one attribute should not depend upon the values assumed by the other attributes.

[3]In his popularization of these results, Peter Sandman has relabeled this axis "outrage." We do not use the term since it has been subject to limited empirical investigation. In addition, high levels of public outrage, at least when measured by calls for government regulatory intervention, may correlate with *both* high levels of dread *and* high levels of uncertainty.

[4]All of the constructed scales suggested in this section are provided as examples. Somewhat different scales might prove preferable in actual implementation.

[5]For example, suppose the best estimate of the expected annual number of deaths is 356 with a 95% confidence interval of ±7. Since the resolution specified for this attribute is a factor of 10, the value can simply be reported as 360 without any indication of uncertainty. On the other hand, if the best estimate of the expected annual number of deaths is 950 with a 95% confidence interval of ±300 then the value should be reported as 950±300. Similarly, when a risk cannot be placed in one of the categories of the constructed subjective scales, this fact should be noted and the range of uncertainty indicated (for example "high, but with about a 10% probability that it could be medium").

[6]We adopted this procedure in a series of small group lay decisionmaking exercises on transmission line siting that we ran at Carnegie Mellon some years ago and found that, in combination with a set of written, well specified tasks the group was to perform, it worked well (Hester and others, 1990).

[7]Note that this voting scheme is intended as simply a heuristic aide to assist participants in rank ordering. While the vote scores will give participants some indication of the strength of feelings that other participants have about different risk categories, no precise meaning should be assigned to the actual weights that result. The overall objective of this exercise is a simple set of ordinal ranks, *not* a ratio-scale measure of relative importance or "utility." While attractive in principle, the latter is too complicated and procedurally ambitious to be used in this exercise.

REFERENCES

Crouch, E.A.C., and R. Wilson. 1981. *Risk/Benefit Analysis.* Cambridge, Massachusetts: Ballinger.

Dryzek, J.S. 1990. *Discursive Democracy.* New York: Cambridge University Press.

Fischhoff, B. 1991. Eliciting Values: Is Anything in There? *American Psychologist* 46: 835–47.

Fischhoff, B., A. Bostrom, and M.J. Quadrel. 1993. Risk perception and communication. *Annual Review of Public Health* 14: 183–203.

Fischhoff, B., P. Slovic, S. Lichtenstein, S. Read, and B. Combs. 1978. How Safe Is Safe Enough?: A Psychometric Study of Attitudes Towards Technological Risks and Benefits. *Policy Sciences* 8: 127–52.

Fischhoff, B., S. Watson, and C. Hope. 1984. Defining Risk. *Policy Sciences* 17: 123–39.

Funtowicz, S.O., and J.R. Ravetz. 1990. *Uncertainty and Quality in Science for Policy*. London: Kluwer Academic Publishers.

Gilovich, T. 1993. *How We Know What Isn't So*. New York: Free Press.

Glickman, Theodore S., and Michael Gough, editors. 1990. *Readings in Risk*. Washington, D.C.: Resources for the Future.

Hester, G., M. Granger Morgan, I. Nair, and H. Keith Florig. 1990. Small Group Studies of Regulatory Decision Making for Power-Frequency Electric and Magnetic Fields. *Risk Analysis* 10: 213–28.

Kahneman, D., P. Slovic, and A. Tversky, editors. 1982. *Judgments under Uncertainty: Heuristics and Biases*. New York: Cambridge University Press.

Keeney, R.L. 1980. *Siting Energy Facilities*. New York: Academic Press.

Keeney, R.L., and H. Raiffa. 1976. *Decisions with Multiple Objectives: Preferences and Value Trade-Offs*. New York: Wiley.

Krimsky, S., and A. Plough. 1988. *Environmental Hazards: Communicating Risks as a Social Process*. Dover, Massachusetts: Auburn House.

Morgan, M. Granger, B. Fischhoff, A. Bostrom, L. Lave, and C.J. Atman. 1992. Communicating Risk to the Public. *Environmental Science and Technology* 26: 2048–56.

Morgan, M. Granger, and M. Henrion. 1991. *Uncertainty*. New York: Cambridge University Press.

NRC (National Research Council). 1981. *Surveys of Subjective Phenomena*. Washington, D.C.: National Research Council.

———. 1989. *Improving Risk Communication*. Washington, D.C.: National Research Council.

Shlyakhter, A.I., and D.M. Kammen. 1993. *Uncertainties in Modeling Low Probability/High Consequence Events: Application to Population Projections and Models of Sea-Level Rise*. New York: IEEE Computer Society Press.

Slovic, P. 1987. Perceptions of Risk. *Science* 236: 280–85.

Slovic, P., B. Fischhoff, and S. Lichtenstein. 1979. Rating the Risks. *Environment* 21(4): 14–20, 36–39.

———. 1980. Facts and Fears: Understanding Perceived Risk. In *Societal Risk Assessment: How Safe Is Safe Enough?* edited by R. Schwing and W. Albers Jr. New York: Plenum Press.

Turner, C., and E. Martin, editors. 1984. *Surveying Subjective Phenomena*. New York: Russell Sage Foundation.

U.S. EPA (Environmental Protection Agency). Office of Policy Analysis. 1987. *Unfinished Business: A Comparative Assessment of Environmental Problems*. Washington, D.C.: U.S. EPA.

Index

Index

Also of Interest from RFF

Readings in Risk

Theodore S. Glickman and Michael Gough, eds.

"A very practical and realistic publication."

—*Chemical and Engineering News*

"Could form the basis for a course in risk analysis. Little mathematical background is required, and each paper is followed by a set of questions for discussion...an excellent text to teach from."

—*American Scientist*

"Compiles the seminal essays on risk issues...presented in a convenient, objective, simple, and stimulating manner...Its organization, selection of papers, and concise but provocative introductory essays make it an understandable and desirable resource for a nontechnical audience...Has its greatest value as a classroom tool."

—*Environmental Science and Technology*

1990 • 262 pages • ISBN 0-915707-55-1 (paper)

Confronting Uncertainty in Risk Management: A Guide for Decision-Makers

Adam M. Finkel

Providing a systematic way to think about, quantify, and respond to uncertainty in risk assessments, this report focuses on the ways in which uncertainty analysis can improve the quality of "routine" risk management actions.

1990 • 87 pages (paper)

Controlling Asbestos in Buildings: An Economic Investigation

Donald N. Dewees

Concerns about the high exposure of workers during installation of asbestos in the past have been widely addressed. The problems posed by asbestos now present in existing buildings, however, are more difficult to deal with. The author develops a methodology for economic analysis of asbestos control programs in existing buildings and presents the results of three case studies.

1986 • 106 pages • ISBN 0-915707-27-6 (paper)

Also of Interest from RFF

Economics and Episodic Disease:
The Benefits of Preventing a Giardiasis Outbreak

Winston Harrington, Alan J. Krupnick, and Walter O. Spofford, Jr.

With exhaustive attention to detail, the authors estimate the social costs to a community arising from an outbreak of waterborne disease. Their appealing blend of economic theory and innovative empirical analysis will help to avoid contaminated drinking water and will enhance the study of food safety issues and public health episodes.

1991 • 202 pages (index) • ISBN 0-915707-59-4 (cloth)

Nuclear Imperatives and Public Trust:
Dealing with Radioactive Waste

Luther J. Carter

"Carter has done a masterful job of laying out the technical issues, the political maneuvering, and the governmental bungling that have occurred during the past three decades of the nuclear-power program."

—Amicus Journal

"Carter presents a detailed and penetrating analysis of the events and policy decisions that led to noncommunist countries' collective failure to manage their civilian nuclear waste problem…This is a valuable book that leaves the reader with a hopeful sense about the future…It is worthwhile reading for both newcomers and veterans of the nuclear debate."

—Chemical and Engineering News

"Refreshingly free of the partisanship that generally clouds the discussions of nuclear power."

—The New York Times Book Review

1987 • 473 pages (index) • ISBN 0-915707-47-0 (paper)

Also of Interest from RFF

Worst Things First? The Debate over Risk-Based National Environmental Priorities

Edited by Adam M. Finkel and Dominic Golding

This book presents findings from a forum convened to explore the controversy over EPA's risk-based approach for setting the nation's environmental priorities. Agreeing that alternative ways exist to target the nation's resources for environmental protection, participants differ sharply as to whether these varied approaches complement each other or would disrupt environmental policy-making.

1994 • 346 pages • ISBN 0-915707-74-8 (cloth) • ISBN 0-915707-76-4 (paper)

Assigning Liability for Superfund Cleanups: An Analysis of Policy Options

Katherine N. Probst and Paul R. Portney

While more than 2,700 emergency removals of hazardous materials have taken place under Superfund, implementing the long-term cleanup program has been the object of considerable controversy. One of the most contentious issues is whether the liability standards in the law should be revised. The authors analyze the pros and cons associated with the current liability scheme and a variety of alternative liability approaches.

"Seems to be setting the agenda for reform of the liability standards under the 1980 Superfund statute."

—*World Insurance Report*

1992 • 62 pages • ISBN 0-915707-64-0 (paper)

Also of Interest from RFF

Footing the Bill for Superfund Cleanups: Who Pays and How?

Katherine N. Probst, Don Fullerton, Robert E. Litan, and Paul R. Portney

The authors look at who pays the costs for cleaning up toxic waste sites under the current Superfund liability scheme on a site-by-site basis. They analyze the incidence of different taxing mechanisms and compare the financial effects on specific industries of the current Superfund program and of several alternative liability and tax mechanisms.

Copublished with the Brookings Institution

1995 • 176 pages • ISBN 0-8157-2994-4 (cloth) • ISBN 0-8157-2995-2 (paper)

The RFF Database of Superfund NPL Sites

This database, developed at RFF and including information on 1,134 sites on EPA's National Priorities List, was created to estimate the effects of different liability schemes on needed trust fund revenues as well as the total transaction costs of the responsible parties and the magnitude of cleanup costs borne by key industry sectors.

1995 • PC-DOS diskette (3.5" and 5.25" high density) • ISBN 0-915707-78-0

Analyzing Superfund: Economics Science, and Law

Edited by Richard L. Revesz and Richard B. Stewart

This book brings together some of the most important theoretical and empirical work from the research community on four issues central to the evaluation of Superfund: cleanup standards, the liability regime, transaction costs, and natural resource damages.

1995 • 263 pages • ISBN 0-915707-75-6 (cloth)